The
Number Devil

The
Number Devil

A Mathematical Adventure

Hans Magnus Enzensberger

Illustrated by
Rotraut Susanne Berner

Translated by
Michael Henry Heim

An Owl Book
Henry Holt and Company ❖ New York

Henry Holt and Company, LLC
Publishers since 1866
115 West 18th Street
New York, New York 10011

Henry Holt® is a registered trademark
of Henry Holt and Company, LLC.

Library of Congress Cataloging-in-Publication Data
Enzensberger, Hans Magnus.
[Zahlenteufel. English]
The number devil / Hans Magnus Enzensberger;
illustrated by Rotraut Susanne Berner.
    p. cm.
  ISBN 0-8050-5770-6
  ISBN 0-8050-6299-8 (pbk.)
  Summary: Annoyed with his math teacher who assigns word
problems and won't let him use a calculator, twelve-year-old
Robert finds help from the number devil in his dreams.
  [1. Mathematics—Fiction. 2. Numbers, Natural—Fiction.
3. Dreams—Fiction.]   I. Berner, Rotraut Susanne, ill. II. Title.
PZ7.E72455Nu   1998   [Fic]—dc21   97-42448

Henry Holt books are available for special promotions and
premiums. For details contact: Director, Special Markets.

Originally published in Germany in 1997 under the title
*Der Zahlenteufel: Ein Kopfkissenbuch für alle,
die Angst vor der Mathematik Haben*

First published in the United States in 1998
by Metropolitan Books

First Owl Books Edition 2000

Printed in Hong Kong
7   9   11   13   15   14   12   10   8   6
13   15   14   12
(pbk.)

For Theresia

The First Night

Robert was tired of dreaming. I always come out looking dumb, he said to himself.

For example, he would dream of being swallowed by a big ugly fish, and even after it was over he could smell the fish's awful stench. Or he'd be sliding down an endless slide, faster and faster, and no matter how many times he cried out *Stop!* or *Help!* on he went until finally he woke up drenched in sweat.

His dreams also played tricks on him whenever he wanted something really bad. Once he had his heart set on a racing bike with twenty-eight gears, and he dreamed that the bike was waiting for him in the basement. It was an unbelievably detailed dream: the bike had a purple metallic finish and was parked next to the wine cabinet. He even knew the sequence of the combination lock: 12345. He couldn't forget *that* now, could he? Well, in the middle of the night, still woozy with sleep, he staggered down to the basement in his pajamas, and what did he find next to the wine cabinet? A dead mouse. That was a low blow!

Eventually Robert came up with a way of dealing with the tricks his dreams played on him. The minute one started, he would think (without waking up), It's just another one of those yucky fish. I know just what's going to happen. It's going to gobble me up. But I also know it's only a dream, because only in dreams can a fish swallow a person. Or he'd think, Here I go sliding again, but there's nothing I can do about it. I can't stop, and I'm not *really* sliding, anyway. And when the fantastic racing bike came back to haunt him, or a computer game he couldn't live without—there it was, right

next to the telephone—he knew it was just a hoax. He didn't even look at the bike; he turned away. But no matter what he did, the dreams kept coming back, and that troubled him.

And then suddenly one night—there was the number devil!

Robert was thrilled to be free of the hungry fish and the endless slide. This time he dreamed of a meadow. The funny thing was that the grass grew so tall that it seemed to reach the sky, or at least over his head. And what did he see but a gigantic beetle glaring at him, a caterpillar perched on a blade of grass, and an elderly man the size of a grasshopper bobbing up and down on a spinach leaf and staring at him with bright and shining eyes.

"Who are you?" Robert asked.

The man responded in a surprisingly loud voice.

"I am the number devil!"

Robert was in no mood to put up with nonsense from a pip-squeak like that.

"First of all, there's no such thing as a number devil."

"Is that so? How can you be speaking to me if I don't exist?"

"And besides . . . I hate everything that has to do with numbers."

"And why is that, may I ask?"

"You sound as though you never went to school. Or maybe you are a teacher yourself?

" 'If 2 pretzel makers can make 444 pretzels in 6

hours, how long does it take 5 pretzel makers to make 88 pretzels?'

"How dumb can you get!" said Robert. "A colossal waste of time if you ask me. So get going! Scram! Shoo!"

But instead of doing as he was bidden, the number devil made an elegant leap and landed smack next to Robert, who was staging a sit-down strike in the tall grass.

"Where does your pretzel tale come from? School, I bet."

"Where else?" said Robert. "Mr. Bockel—he's our teacher, a new teacher—well, he's always hungry, though he's got plenty of fat on him. Whenever he thinks we're not looking because we're so into the problems he gives us, he sneaks these pretzels out of his briefcase and wolfs them down."

"I see," said the number devil with a wry smile. "I have nothing against your Mr. Bockel, but that kind of problem has nothing whatever to do with what I'm interested in. Do you want to know something? Most genuine mathematicians are bad at sums. Besides, they have no time to waste on them. That's what pocket calculators are for. I assume you have one."

"Sure, but we're not allowed to use them in school."

What did Robert see but an elderly man the size of a grasshopper bobbing up and down on a spinach leaf and staring at him with bright and shining eyes.

"I see," said the number devil. "That's all right. There's nothing wrong with a little addition and subtraction. You never know when your battery will die on you. But *mathematics*, my boy, that's something else again!"

"You're just trying to win me over," said Robert. "I don't trust you. If you give me homework in my dream, I'll scream bloody murder. That's child abuse!"

"If I'd known you were going to be such a scaredy-cat, I wouldn't have entered your dream. All I want is to perk you up a little, and since I'm off duty most nights, I thought I'd spare you those endless slides you've been going down."

"Gosh, thanks."

"I'm glad you understand."

"But I hope *you* understand that I won't let you take me for a ride."

Suddenly the number devil leaped up out of the grass, a pip-squeak no more.

"That's no way to talk to a devil," he shouted, his eyes sparkling, and he trampled the grass until it was all flat.

"I'm sorry," Robert said meekly, though the whole thing was getting weirder and weirder. "But if talking about numbers is as simple as talking about movies or bikes, why do they need their own devil?"

"You've hit the nail on the head, my boy," the devil replied. "The thing that makes numbers so devilish is precisely that they *are* simple. And you don't need a calculator to prove it. You need one thing and one thing only: one. With one—I am speaking of the numeral, of course—you can do almost anything. If you are afraid of large numbers—let's say five million seven hundred and twenty-three thousand eight hundred and twelve—all you have to do is start with

$$1 + 1$$
$$1 + 1 + 1$$
$$1 + 1 + 1 + 1$$
$$1 + 1 + 1 + 1 + 1$$
$$\cdots$$

and go on until you come to five million etcetera. You can't tell me that's too complicated for you, can you? Any idiot can see that."

"Right," said Robert.

"And that's not all," the number devil added, picking up a walking stick with a silver knob and twirling it in front of Robert's nose. "When you get to five million etcetera, you can go on. Indefinitely. There's an infinite number of numbers."

Robert didn't quite know whether to believe him.

"How can you be so sure?" he asked. "Have you ever tried?"

"Can't say I have," the devil answered. "It would take too long, for one thing. And it makes no sense anyway. It would be a waste of time."

Robert didn't see why.

"Either I can count to the end, in which case there is no such thing as infinity, or there is no end and I can't count to it."

"Wrong!" the number devil shouted, his mustache quivering, his eyes bulging, and his face turning red with rage.

"What do you mean 'wrong'?" Robert asked.

"You nincompoop! Tell me, how many pieces of chewing gum do you think have been chewed to this day?"

"I have no idea."

"Guess."

"Billions," said Robert. "If you take my friends Al and Betsy and Charlie, and the rest of the kids in the class, and in the city, and the country, and the world . . . trillions!"

"At least," said the number devil. "Okay, now

let's pretend that everyone's gone on a chewing spree and we're down to the last piece of chewing gum. I pull another one out of my pocket, the last one that I've saved for myself, and what have we got? All those trillions of chewed pieces of chewing gum plus one. Do you see what I mean? I don't really *need* to count them. All I need is a recipe to take care of anything that comes along. And that I have."

After thinking over all that the number devil had said, Robert was forced to admit that he had a point.

"By the way, the reverse is true as well," the number devil added.

"The reverse? What do you mean?"

"Simple," said the number devil with a grin. "Just as there are infinitely large numbers, there are infinitely small numbers. And an infinite number of infinitely small numbers." And so saying, he twirled his walking stick like a propeller in Robert's face.

I'm starting to feel dizzy, thought Robert. It was the same feeling he'd had on the slide.

"Stop!" he shouted.

"Why so jittery, Robert?" asked the number devil. "It's perfectly harmless. Look. I just pull another piece of chewing gum out of my pocket . . ."

And he did. Only it was as long as a ruler, hard as a rock, and a strange shade of purple.

"You call that chewing gum?" Robert asked.

"The chewing gum of your dreams," the number devil replied. "And I'm going to share it with you. Watch carefully now. As long as it's whole, it's *my* chewing gum. One piece, one person." And sticking a piece of chalk—also a strange shade of purple—to the end of his walking stick, he said, "Here's how we write it—"

He traced the two ones on the sky, the way planes skywrite advertisements. The purple numbers floated for a while against a bank of white clouds, then melted slowly like a scoop of raspberry ice cream.

"Cool!" Robert said. "What I wouldn't give for a stick like that!"

"Oh, it's nothing special. Though it does write

on anything: clouds, walls, screens. And I never need a notebook or briefcase. But that's neither here nor there. Let's get back to our chewing gum. If I break it in two, you have a half and I have a half. One gum, two people. The gum goes on top, the people on the bottom:

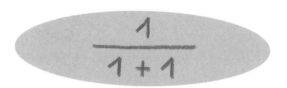

$$\frac{1}{1+1}$$

Now your friends will want some too, of course."

"Al and Betsy."

"As you like. And let's say Al asks you for gum, and Betsy asks me, and we share equally. That means we each get a quarter:

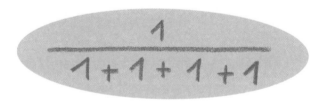

$$\frac{1}{1+1+1+1}$$

But that's just the beginning. Everyone else will want a piece. The rest of the class, the rest of the city . . . So each of us four will have to break his fourth in half, and then that will have to be halved, and that halved, and that halved, and so on."

"Till the cows come home," said Robert.

"Till the pieces are so tiny that you can't see them with the naked eye. Be that as it may, we go on halving them until the six billion people on earth have all had their share. But then come the six hundred billion mice, and they all want *theirs*. In other words, we'll never come to the end of it."

Meanwhile he had been writing an endless row of purple ones along the sky.

"You'll fill the whole world if you go on like this!" Robert cried.

"Well," said the number devil, puffing himself up again, "I'm only doing it for you! You're the one who's afraid of numbers. You're the one who wants everything simple so you won't get mixed up."

"But all those ones get so boring after a while. I don't think it's simple. It's just that all those ones give me a headache. They actually make things more complicated than they are."

"Well, well," said the number devil, clearing the sky with a casual wave of the hand. "So you agree we need something less clumsy than $1+1+1+1 \ldots$ Numbers, for instance. Which is why I invented them."

"You? You expect me to believe that you invented numbers?"

"Me or a few others. It doesn't matter who exactly. Why are you so suspicious? What do you say I show you how to make all numbers out of ones?"

"Okay. How?"

"Simple. I start as follows:

And go on to:

I bet you need your calculator for that."

"Don't be silly," said Robert.

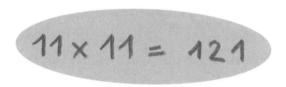

"See?" said the number devil. "You've made a two out of nothing but ones. Now try this—"

$$111 \times 111$$

"That's too hard. I can't do it in my head."

"Then use your calculator."

"My calculator! You don't think I take it to bed with me, do you?"

"Then use this one," he said, pressing one into Robert's hand. It had a funny feel to it, slimy, like dough, and it was a sickly shade of green, but it worked. Robert entered:

and got:

$$12321$$

"Cool," said Robert. "Now we have a three."
"Right. Just keep going."
So Robert entered the following:

$$1111 \times 1111 = 1234321$$
$$11111 \times 11111 = 123454321$$

"Very good," said the number devil, patting Robert on the shoulder. "I'm sure you've noticed that not only do you get a new number each time, you get a number that reads the same forward and backward, like ANNA or TOOT or ROTATOR."

Robert thought that this was a pretty good trick, and so he tried six ones as well—and it worked! But when he got to

the calculator gave up the ghost. To Robert's surprise, it suddenly went *Pfft!* and melted down into a sickly green goo.

"Yuck!" said Robert, wiping the green mess from his fingers.

"All you need is a bigger calculator or a computer. A computer would spit out the answer in no time."

"Are you sure?"

"Of course I'm sure."

Robert thought that the number devil was a bit too confident. Maybe he was just bluffing. Robert decided to take a chance and said, "You haven't tried it with

$$11\ 111\ 111\ 111 \times 11\ 111\ 111\ 111$$

have you?"

"No, can't say I have."

"Well, I bet it doesn't work."

The number devil started doing the problem in his head, but his face turned bright red again and swelled up like a balloon. Was it because he was angry, Robert wondered, or because the problem was hard?

"Wait a second," the number devil mumbled. "I can't seem to come up with anything. Damnation! You were right. It doesn't work. How did you know?"

"I didn't. You don't think I'm crazy enough to try a problem like that do you? I was just guessing."

"Guessing? Guessing is not allowed in mathematics! Mathematics is an exact science!"

"But when you said that numbers don't stop, that they go on till the cows come home, that was a guess, wasn't it?"

"How dare you? What are you, anyway? A beginner! A rank amateur! And you want to teach me my trade?"

He grew bigger and fatter with every word; he started huffing and puffing. Robert was frightened.

"You pinhead! You pip-squeak! You stuck-up

little number midget!" he screamed, and no sooner had the number devil got the last word out than he burst with a great bang.

Robert woke up. He had fallen out of bed and was a little dizzy, but he laughed to think he had outwitted the number devil.

The Second Night

Robert was sliding. The same old story. It had
started the minute his head hit the pillow, and he
couldn't stop. This time he was sliding along a
tree. Don't look down, he thought, clinging to the
tree for dear life and sliding on and on . . .

Then all at once he landed with a plop on a soft
bed of moss. He heard a giggle and who did he see
but the number devil perched on a velvety brown
mushroom, smaller than Robert had remembered,
and staring down at him with his shining eyes.

"How did *you* get here?" he asked Robert.

Robert pointed to the tree trunk, which stretched
as far as the eye could see. But it was not alone: there
was a whole forest of them. And they weren't so
much trees as ones. He had landed in a forest of
ones.

But that was not all. The air hummed with tiny
flylike numbers dancing in front of his nose. He
tried shooing them away, but there were too many:

twos, threes, fours, fives, sixes, sevens, eights, and nines kept brushing against him. Robert had always hated moths and gnats and couldn't stand having the beasties around.

"Are they bothering you?" asked the number devil. He put his hand to his mouth and blew them all away with a *Pfft!* Suddenly the air was clear of everything but the forest of ones reaching up to the sky.

"Have a seat, Robert," said the number devil.

Robert was surprised to find him so friendly.

"Where? On a mushroom?"

"Why not?"

"But I'd feel silly. Where are we, anyway? In a picture book? Last time you sat on a spinach leaf, and now you're on a mushroom. I seem to remember something like that. In a book I once read."

"The mushroom in *Alice in Wonderland* perhaps."

"But what's the connection between a made-up story and mathematics?"

"The kind of connection you make when you dream, my boy. You don't think *I* was behind all those flies, do you? No, I'm wide awake. You're the one in bed dreaming. Now what do you say? Are you going to stand there forever?"

Don't look down, Robert thought, clinging to the tree for dear life and sliding on and on . . . He had landed in a forest of ones.

Robert saw he had to do something, so he clambered onto the nearest mushroom. It was enormous and, except for a few bumps, as soft and cozy as an armchair.

"How do you like it?"

"It's fine," said Robert. "I just wonder who came up with the number flies and the forest of ones. I couldn't have. Not in my wildest dreams. It could only have been you."

"And if it was?" said the number devil, preening himself on his mushroom. "Though there's still something missing."

"What?"

"Nothing. I mean, zero."

He was right. There hadn't been a single zero among all the flies.

"Why?"

"Because zero was the last number to be discovered. Which isn't surprising, given that zero is the most sophisticated of numbers. Here, look." And finding a space between two tree-high ones, he wrote some letters in the sky with his walking stick:

$$MCM$$

"Tell me, when were you born, Robert?"

"Me? In 1986," said Robert a bit reluctantly.

MCMLXXXVI

wrote the number devil.

"I know what those are. Those are those old numbers you sometimes find in the cemetery."

"And they come from the ancient Romans. Who had a tough time of it, by the way. Partly because their numbers were so hard to read. Though this one is easy enough—"

I

"One," said Robert.

"And this?"

X

"X is ten."

"Right. And this, my boy, is the year you were born:

MCMLXXXVI

The first M means 1,000. C is 100, but because it comes before the second M you must subtract it to get 900. L is 50 and X is 10. You add them together, which gives you 80. V is 5, which you add to our friend 1 to get 6. So this is 1,000 + 900 + 80 + 6."

"Gosh, that's awfully complicated."

"Right again. And you know why? Because the Romans had no zero."

"I don't see the connection. Besides, what's so great about zero? Zero means nothing."

"Which is precisely what is so brilliant about it."

"But why even call it a number? Nothing doesn't count."

"Don't be so sure. Remember how we divided up that piece of chewing gum among all those billions of people and hundreds of billions of mice? And how the portions got so small that in the end you couldn't see them, not even with a microscope? Well, we could have gone on forever without reaching zero. We'd have come closer and closer, but we'd never have made it."

"So?"

"So we've got to try something else. Minus, for instance. Yes, that should do the trick."

He stretched out his walking stick and tapped the end of one of the ones. It shrank and shrank until it stood meek and manageable at Robert's feet.

"Go at it," said the number devil.

"What do you mean?"

"Try the minus."

$$1 - 1 =$$

"One minus one is zero," said Robert. "Everyone knows that."

"You see? You see how necessary zero is? You can't do without it."

"But why do we need to write it? If nothing is left, why not just leave a blank? Why invent a number for something that doesn't exist?"

"Try this, then."

$$1 - 2 =$$

"That's easy," Robert said. "One minus two is minus one."

"Right. But look what you get without a zero.

$$\ldots 4, 3, 2, 1, -1, -2, -3, -4 \ldots$$

The difference between four and three is one. Between three and two—one. Between two and one—one. And between one and minus one?"

"Two."

"Which means there must be a number missing between one and minus one."

"That tricky little zero."

"I told you we couldn't do without it. That brings us back to the Romans. They reckoned they could, and look what happened. Instead of 1986, they had to fiddle with all those M's, C's, L's, X's, and V's. The Romans had to give each number a different letter, all because they didn't have zero."

"But what's that got to do with our chewing gum and the minus?"

"Forget the chewing gum. Forget the minus. The zero's real beauty lies elsewhere. But you'll need to use your head to appreciate it. Are you up for it or are you too tired?"

"No, as long as I'm not sliding I'm fine. In fact, I like it here on the mushroom."

"Good. Then let me give you a little problem to solve."

Why is he suddenly being so nice to me? Robert wondered. I bet he's got something up his sleeve. But all he said was "Fire away."

And the number devil asked:

$$9 + 1 =$$

"Ten," Robert answered like a shot. "Is that all?"

"And how do you write it?"

"I haven't got a pen."

"Then skywrite it. Here, take my walking stick."

$$9 + 1 = 10$$

Robert wrote in purple cloud script.

"One and zero?" the number devil said. "One plus zero doesn't equal ten."

"Oh, come off it!" Robert shouted. "I didn't write one plus zero! I wrote a one and a zero, and that's ten!"

"And why, may I ask, is that ten?"

"Because that's the way you write it."

"And why do you write it that way?"

"Why, why, why?" Robert moaned. "You're getting on my nerves."

"Want me to tell you?" the number devil asked, leaning back leisurely on his mushroom.

A long silence followed.

"Fine! Go ahead!" Robert blurted out when he couldn't bear it any longer.

"Simple. It comes from hopping."

"Hopping?" Robert said scornfully. "What's that supposed to mean? Numbers don't hop."

"Numbers hop if *I* tell them to hop," the number devil replied. "Don't forget who you're dealing with."

"All right, all right," Robert said. "Just tell me what you mean by hopping."

"Gladly. Let's go back to square one. Or, rather, the number one.

$$1 \times 1 = 1$$
$$1 \times 1 \times 1 = 1$$
$$1 \times 1 \times 1 \times 1 = 1$$

Tack on as many ones as you like and you still get one for your answer."

"Sure. But what's your point?"

"You'll see if you try the same thing with two."

"Okay," said Robert.

$$2 \times 2 = 4$$
$$2 \times 2 \times 2 = 8$$
$$2 \times 2 \times 2 \times 2 = 16$$
$$2 \times 2 \times 2 \times 2 \times 2 = 32$$

. . .

"Wow, that goes fast! If I go much further, I'll need my calculator."

"It goes even faster if you start with five. Why don't you give it a try?"

$$5 \times 5 = 25$$
$$5 \times 5 \times 5 = 125$$
$$5 \times 5 \times 5 \times 5 = 625$$
$$5 \times 5 \times 5 \times 5 \times 5 = 3125$$
$$5 \times 5 \times 5 \times 5 \times 5 \times 5 = 15625$$

"Whoa!" Robert shouted.

"Why do large numbers make you so jumpy? I can assure you that most large numbers are perfectly harmless."

"Says you!" said Robert. "Besides, I don't see the point of multiplying five by itself over and over."

"I'm coming to that. You know what the number devil does instead of writing all those boring fives? He writes:

$$5^1 = 5$$
$$5^2 = 25$$
$$5^3 = 125$$

and so on. Five to the first, five to the second, five to the third. In other words, I make the numbers hop.

"Now do you see? Do the same with ten, and it's as easy as pie. You can throw your calculator away. Make the ten do one hop, and it remains exactly as is:

$$10^1 = 10$$

Make it hop twice, and you get:

$$10^2 = 100$$

Make it hop three times, and you get:

$$10^3 = 1000$$

"So if I make it hop five times," Robert cried, "I get 100,000! Once more, and I get a million!"

"And so on, till the cows come home," said the number devil. "Simple, eh? That's the beauty of the zero. It lets you hold a space and move on. You can always tell a number's value by its position: the farther to the left it is, the more it's worth; the farther to the right, the less.

"When you write 555, you know the last five is worth exactly five and no more; the next-to-last five is worth ten times more—that is, fifty; and the first five is worth a hundred times the last one— that is, five hundred. And why? Because it's been hopped up front.

"Now, the fives of the ancient Romans could never be anything but fives. Why? Because the Romans didn't know how to hop. And why didn't they know how to hop? Because they had no zero to keep places. Which meant they ended up with monstrosities like MCMLXXXVI.

"So rejoice, my boy! You're much better off than the Romans. With the help of friend zero and a bit of hopping you can produce any number, big or small, any number you please. 786, for instance."

"But I don't need 786."

"Really now! You're brighter than that! Try the year you were born, 1986." The number devil started growing again, and his mushroom followed suit. "Well, what are you waiting for?" he bellowed. "Get a move on!"

There he goes again, thought Robert. Get him worked up about something and he's impossible. Worse than Mr. Bockel.

He carefully wrote a large one in the sky.

"Wrong!" the number devil screamed. "Dead wrong! How did I wind up with a fool like you? I told you to *produce* the number, not scribble it down."

Robert would have given anything to wake up. I'm not going to put up with this, he thought, watching the number devil's head swell up and turn redder and redder.

"The end," the number devil called down to him.

Robert stared back, completely at a loss.

"Start at the end, not the beginning."

"If you say so," said Robert, who was in no mood to argue. He erased the one and wrote a six in its place.

"Finally caught on, have you? Well then, we may proceed."

"No problem," said Robert warily, "though I'd appreciate it if you didn't fly off the handle over every detail."

"Sorry, but what do you expect? I'm the number devil, not Santa Claus."

"How do you like my six?"

The number devil shook his head and wrote:

$$6 \times 1 = 6$$

"But that's the same," Robert protested.

"You'll see what I have in mind. Now comes the eight. And don't forget to hop."

Suddenly Robert did see what he had in mind. He wrote:

$$8 \times 10 = 80$$

"I get it!" he shouted before the number devil could tell him what to do. "With nine I make the ten hop twice." And he wrote:

$$9 \times 100 = 900$$

and

$$1 \times 1000 = 1000$$

"That was a triple hop," he said. "And now for the grand total—

$$6 + 80 + 900 + 1000 = 1986$$

It's not so hard after all. I don't even need the number devil."

"So you don't need the number devil, eh? You're getting too big for your breeches, boy! All you've had so far are the most ordinary numbers. Nothing to write home about. Wait till I start pulling numbers out of my hat, all kinds of numbers—unreal numbers, unreasonable numbers. You have no idea how many kinds of numbers there are! Numbers that run around in circles, numbers that play tricks with your brain, numbers that go on forever . . ."

As he spoke, his grin grew wider and wider. Robert could see all the teeth in his mouth; they too seemed to go on forever. And then he started twirling his walking stick in front of Robert's face again.

"Help!" Robert screamed, and woke up.

As the number devil spoke, his grin grew wider and wider. Robert could see all the teeth in his mouth; they too seemed to go on forever.

Still in a daze the next morning, Robert said to his mother, "Do you know the year I was born? It was 6×1 and 8×10 and 9×100 and 1×1000."

"I don't know what's got into the boy lately," said Robert's mother, shaking her head. "Here," she added, handing him a cup of hot chocolate, "maybe this will help. You say the oddest things."

Robert drank his hot chocolate in silence. There are some things you can't tell your mother, he thought.

The Third Night

In time Robert grew accustomed to dreaming of the number devil. He even came to look forward to it. True, he could have done without his know-it-all attitude and his temper tantrums—you could never tell when he'd blow up and yell at you—but it was better, so much better, than being swallowed by a slimy fish or sliding down and down into a black hole.

Besides, Robert had made up his mind to show the number devil that he was no fool. You have to put people like him in their place, Robert thought as he got ready for bed one night. The big ideas he has about himself—and all because of a zero. *He* wasn't much more than a zero when you got down to it. All you had to do was wake up and he was gone.

But to put him in his place Robert had to dream of him, and to dream of him he had to fall asleep. And

Robert suddenly noticed he was having trouble doing so. For the first time in his life he lay awake in bed, tossing and turning.

"What are you tossing and turning for?"

All at once, Robert realized his bed was in a cave. There were weird paintings of animals on the stone walls, but he had no time to study them because the number devil was standing over him, twirling his walking stick.

"Rise and shine, Robert!" he said. "Today's our division day."

"Must I?" Robert asked. "You might have at least waited until I was asleep. Besides, I hate division."

"Why?"

"When you add or subtract or even multiply, things come out even. What bugs me about division is that you get this remainder."

"The question is when."

"When what?"

"When you get a remainder and when you don't. That's what counts. You can tell just by looking at them that some numbers can be divided evenly."

"Right. Like even numbers, which can all be divided by two. No problem. I'm pretty good at threes as well:

$$9 \div 3$$
$$15 \div 3$$

and so on. It's like multiplying in reverse:

$$3 \times 5 = 15$$

becomes

$$15 \div 3 = 5$$

I don't need a number devil for that. I can do it on my own."

Robert shouldn't have said that. The number devil, his mustache quivering, his nose reddening, his head growing bigger and bigger, jerked Robert out of bed.

"What do *you* know?" the number devil shouted. "Just because you've learned the multiplication table you think you know all there is to know. Well, you know nothing! Nothing whatsoever!"

There he goes again, thought Robert. First he drags me out of bed, then he hits the ceiling when I tell him I can do division.

All things being equal, Robert would have upped and left, but how do you up and leave a dream? He looked all over the cave, but could find no way to leave.

"Here I come to a rank beginner out of the goodness of my heart, and no sooner do I open my mouth than he starts making wisecracks!"

"The goodness of your heart!" Robert cried. All things being equal, he would have upped and left, but how do you up and leave a dream? He looked all over the cave, but could find no way to leave.

"What are you looking for?"

"A way out."

"If you go now, you'll never see me again! I'll leave you to choke on Mr. Bockel's pretzel problems, or die of boredom in his class."

Robert knew when he was licked.

"I apologize," he said. "I didn't mean to offend you."

"Good," said the number devil, his anger subsiding as quickly as it had come. "Now, nineteen. Try nineteen. See if you can divide it without a remainder."

Robert thought and thought.

"The only way I can come up with," he said at last, "is to divide it by nineteen. Or into nineteen equal parts."

"Doesn't count," the number devil replied. "It's too easy."

"Or divide it by zero."

"Out of the question."

"Out of the question? Why?"

"Because it's forbidden. Dividing by zero is strictly forbidden."

"What if I did it anyway?"

"Then all mathematics would come apart at the seams!"

He was about to lose his temper again, but he managed to pull himself together.

"Tell me," said the number devil, "what would you get if you divided nineteen by zero?"

"I don't know. A hundred, maybe. Or zero. Or anything in between."

"But didn't you say when you were talking about the threes that division was like multiplying in reverse? If that's the case, then

$$3 \times 5 = 15$$

means that

$$15 \div 3 = 5$$

Well, now try that with nineteen and zero."

"Nineteen divided by zero is, say, 19."

"And in reverse?"

"19 times zero . . . 19 times zero . . . is zero."

"You see? And no matter what number you take, you always get zero. Which means you must never divide a number by zero."

"Okay," said Robert, "I give up. But what do we do with the nineteen? No matter what number I divide it by—two, three, four, five, six, seven, eight, or nine—I get stuck with a remainder."

"Come a little closer," said the number devil to Robert, "and I'll tell you a secret."

Robert leaned so close to the number devil that his mustache tickled his ear.

"There are two types of numbers," he whispered. "The garden variety, which can be divided evenly,

and the rest, which cannot. I much prefer the latter. You know why? Because they're such prima donnas. From the very first they've caused mathematicians no end of trouble. Wonderful numbers those! Like eleven, thirteen, or seventeen."

Robert couldn't get over how blissful the number devil looked. He might have had a piece of chocolate melting in his mouth.

"Now tell me, dear boy, what the first prima-donna numbers are."

"Zero," said Robert, to get his dander up.

"Zero is forbidden!" the number devil shouted, brandishing his walking stick.

"All right, then. One."

"One doesn't count. How many times do I have to tell you?"

"Okay, okay," said Robert. "Calm down! Two, to begin with. Because, like all prima donnas, it can only be divided by one and itself. And three—or at least I think so. Not four. We've been through that. Five for sure; five isn't divisible by anything. And . . . and so on."

"Ha! What is that supposed to mean?" he said, rubbing his hands together, a sure sign he had something up his sleeve. "The wonderful thing about prima donnas is that no one knows their 'and so on.' No one but me, of course. And I won't tell."

"Not even your friends?"

"Nobody! Never! The thing is, you can't know whether a number is a prima donna or not merely by looking at it. No ordinary mortal can know without testing it."

"And how do you test it?"

"You'll see," he said, and started scribbling all over the wall of the cave with his walking stick.

This is what it looked like when he was through:

	2	3	4	5	6	7	8	9	10
11	12	13	14	15	16	17	18	19	20
21	22	23	24	25	26	27	28	29	30
31	32	33	34	35	36	37	38	39	40
41	42	43	44	45	46	47	48	49	50

"Now take my stick, young man, and tap every number that isn't a prima donna. That will make it disappear."

"But there's no zero," Robert complained, "no one."

"How many times must I tell you! Zero and one are unlike all other numbers: they are *neither* prima donnas *nor* the ordinary kind. Don't you remember what you dreamed back at the very beginning? That all numbers come from one? And then later we saw the need for zero."

"Whatever you say," said Robert. "Anyway, first I'll tap the even numbers, because they're all divisible by two. That's easy."

"All the even numbers but two," the number devil warned him. "Don't forget: two is a prima-donna number."

Robert picked up the stick and in no time flat the wall looked like this:

	2	3		5		7		9	
11		13		15		17		19	
21		23		25		27		29	
31		33		35		37		39	
41		43		45		47		49	

"Now do three. Three is a prima-donna number, but the rest of the numbers in the three column of the multiplication table—six, nine, twelve, etcetera—are not, because they can be divided by three."

When Robert had taken care of three, the following numbers were left:

	2	3		5		7		
11		13				17		19
		23		25				29
31				35		37		
41		43				47		49

"Now four," said Robert. "No, no. We don't need to bother about numbers divisible by four: they're all gone, because four isn't a prima donna, four is 2 × 2. But five, five is a prima donna. Not ten, though, which is gone because it's 2 × 5."

"And you can tap all the ones that end in five."

"Right."

	2	3		5		7		
11		13				17		19
		23						29
31						37		
41		43				47		49

Robert was hitting his stride.

"We can forget about six—six is 2×3—but seven is a prima-donna number."

"Good for you!" the number devil cried.

"Eleven too."

"And what others?"

That, my dear readers, is a question you yourselves must find the answer to. Copy the chart and go on tapping until only prima-donna numbers are left. Let me give you a hint: There are exactly fifteen of them in the chart—no more and no less.

"Well done, Robert," said the number devil, lighting his pipe and chuckling to himself.

"What's so funny?" Robert asked.

"It's not so hard if you stop at fifty," he answered with a wicked grin and settling into a comfortable cross-legged position. "But what if you have a number like

$$10\ 000\ 019$$

or

141 421 356 237 307

Is it a prima donna or isn't it? If you knew how many mathematicians have racked their brains over the issue. Why, even the greatest number devils have come to grief over it."

"But I thought you said *you* knew the 'and so on' of prima-donna numbers. You just didn't want to let me in on it."

"Well, maybe I overstated my case a little."

"I'm glad you can admit you're not perfect," said Robert. "Sometimes you sound less like a number devil than a number dictator."

"The more simpleminded number devils use giant computers. They keep them running for months at a stretch. The trick I taught you—taking care of the twos and threes and fives first—is old hat really. Not that it doesn't work, but when the numbers start getting really big, you know that there's no end to them. We can make them grow bigger than the universe just by adding, multiplying, and hopping. Now there are all kinds of more sophisticated ways of doing things, but clever as they are they don't seem to get us very far. That's what makes them so devilishly interesting—and what is devilish is fun, don't you think?"

And so saying, he twirled his stick with great relish.

"Yes, but what's the point of it all?" Robert asked.

"You *do* ask stupid questions! The world of numbers is never so musty as your Mr. Bockel—Mr. Pretzel—makes it out to be. Luckily you've got me to initiate you into some of the secrets. Like this, for instance: Take any number larger than one and multiply it by two."

"222?" said Robert. "Times two is 444."

"Between the first and the second number there is always—and when I say always, I mean *always*—at least one prima donna."

"Are you sure?"

"307," said the number devil. "But it works with gigantic numbers too."

"Where do you learn these things?"

"You haven't seen anything yet!" he said, savoring Robert's curiosity. Nothing could stop him now. "Take any even number—any one at all, so long as it's larger than two—and I can find two prima donnas that add up to it."

"Forty-eight," Robert said.

"Thirty-one plus seventeen," said the number devil without blinking an eye.

"Thirty-four," said Robert.

"Twenty-nine plus five," said the number devil, without even taking the pipe out of his mouth.

"And it works all the time?" Robert asked, amazed. "Why? How come?"

"To tell you the truth," the number devil said, his forehead wrinkling and his eyes looking upward at the smoke rings he was blowing, "I wish I knew. Nearly every number devil of my acquaintance has tried to come up with an explanation. It always works, but no one knows why."

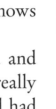

Pretty nifty, thought Robert. He laughed and said, "Well, I think it's great." But what he really thought was great was that the number devil had told him the inside story.

For a while the number devil had a rather crotchety expression on his face (as he always had when he didn't know quite what to do), but when he heard Robert laugh he laughed too and started puffing on his pipe again.

"You're not so stupid as you look, my boy. I'm sorry I have to leave you, but I've got a few more mathematicians to visit tonight. I get a kick out of torturing them a tad."

With that, the number devil began to grow thin. No, not so much thin as transparent. And suddenly all that was left of him was a cloud of smoke and the scribbling on the wall. Then the wall started swimming before Robert's eyes and the cave felt as soft and warm as a blanket. Robert tried hard to remember what was so wonderful about prima-donna numbers, but his thoughts were all white and cloudy like a cotton mountain.

He had rarely slept so well.

And you? Let me show you one last trick—if you haven't dozed off, that is. It works with odd as well as even numbers. Think of a number, any number, so long as it's bigger than five. Fifty-five, say, or twenty-seven.

You can find prima-donna numbers that add up to them too, only instead of two you'll need three. Let's use fifty-five as our example.

$$55 = 5 + 19 + 31$$

Now try twenty-seven. It always works—you'll see—though I can't explain why.

The Fourth Night

"The places you drag me to! A cave with no opening, a forest with ones for trees and mushrooms the size of armchairs. What about today? Where am I anyway?"

"At the beach. Can't you tell?"

Robert looked around: white sand far and wide, the number devil perched on an overturned rowboat, the surf rolling in behind him, and not a soul in sight.

"I bet you've forgotten your calculator again," the number devil said.

"Look, how many times do I have to tell you? I can't take all my stuff to bed with me at night. Do *you* know what you're going to dream the night before you dream it?"

"Of course not," the number devil answered. "Still, if you dream of me, you can just as easily dream of your calculator. But no, I've got to come up with one by magic! I've got to do everything for you! And then you complain it's too soft or too green or too sticky!"

"It was better than nothing."

"All right. I'll try," the number devil said. He raised his wand, and a new calculator appeared before Robert's eyes. If the last one was like a frog, this one was like a soft, furry bed or sofa. And gigantic. At one end it had a little board with fur-covered number keys, but the screen for the numbers stretched the entire length of the backrest. It was one weird calculator.

"Now enter one divided by three," the number devil ordered, "and see what you come up with."

$$1 \div 3$$

Robert punched the keys. The following answer came up in green on the long, long screen:

$$0,3333333333333333333$$

"Doesn't it ever stop?" Robert asked.

"Of course it does," the number devil answered. "It stops where the screen stops."

"And then what?"

"Then it goes on. You just can't see it."

"But it's always the same, one three after the next. How boring!"

"Right."

"And dumb too! It's much easier to write one-third:

$$\frac{1}{3}$$

Then I don't have to worry about all those threes creeping up."

"True," said the number devil, "but then you've got to deal with fractions, and fractions, if I'm not mistaken, are something you can't abide. 'If ⅓ of 33 bakers can make 89 pretzels in 2½ hours, then how many pretzels can 5¾ bakers make in 1½ hours?' "

"No! No! Anything but that! Give me decimals any time! Even if the numbers never end. I'd just like to know what all those threes are doing there."

"Simple. The first three after the dot means three-tenths. The second means three-hundredths, the third three-thousandths, and so on. You can take it from there on your own.

$$0,3$$
$$0,03$$
$$0,003$$
$$0,0003$$
$$0,00003$$
$$\ldots$$

Get it? Good. Then try multiplying everything by three: the three, the three-tenths, the three-hundredths, and so on."

"No problem," said Robert. "I can do that in my head."

$$0,3 \times 3 = 0,9$$
$$0,03 \times 3 = 0,09$$
$$0,003 \times 3 = 0,009$$
$$0,0003 \times 3 = 0,0009$$

"Good. Now what happens if you add all the nines together?"

"Let's see: 0.9 + 0.09 = 0.99, and 0.99 + 0.009 = 0.999. Nines down the line. I bet it could keep on like that forever."

"Right you are. Though if you think about it, there's something fishy going on: ⅓ + ⅓ + ⅓ = 1, doesn't it? Because a third multiplied by three equals a whole. Always has and always will. Well? What do you think?"

"I don't know," said Robert. "Something is still missing—0.999 is *nearly* one, but it doesn't quite get there."

"That's the point. That's why you've got to keep the nines going and never stop."

"Easier said than done."

"Not for a number devil."

With another little chuckle he waved his walking stick and in the twinkling of an eye the sky was filled with an endless chain of purple nines slithering higher and higher.

"Stop!" Robert shouted. "Please stop! It's making me sick."

"A snap of my fingers and they're gone, but not until you admit that the chain of nines behind the zero, if it goes on forever, will turn out to be equal to one."

Meanwhile the chain had kept growing and the sky slowly darkened with nines. Robert was now as dizzy as he was nauseous, but he refused to give in.

"Not on your life!" Robert shouted. "No matter

The number devil waved his walking stick and in the twinkling of an eye the sky was filled with an endless chain of purple nines slithering higher and higher.

how many nines you add to your chain, there will always be something missing: the last nine."

"There is no last nine!" the number devil furiously shouted back.

Robert no longer jumped out of his skin each time the number devil lost his temper. By now he knew that whenever it happened there was something interesting coming up, something the number devil couldn't easily explain. But the chain was flapping dangerously close to Robert's head and had wound so tightly round the number devil that much of him had receded from view.

"All right," Robert said. "I give in. But only if you get rid of the nines."

"It's about time," said the number devil, raising his stick, which now had several layers of nines entwined around it. Then he mumbled some gibberish to himself and the chains disappeared in a flash.

"Phew!" said Robert. "Does it only happen with threes and nines, or do other numbers make such awful chains too?"

"There are chains as endless as sand on the beach. How many would you say there are between 0.0 and 1.0?"

Robert thought long and hard.

"An infinite number. As many as between one and till the cows come home."

"Not bad," said the number devil. "Quite good, in fact. But can you prove it?"

"I can."

"Show me."

"All I have to do is write a zero and a dot," said Robert, "and a one after the dot: 0.1. Then a two after that. And so on. If I keep going, I'll have put all the numbers that have ever been written after the dot before I come to 0.2.

"Whole numbers all."

"Of course. All whole numbers. Every number between one and bezillion can have a zero and a dot before it, and every one of them is less than one."

"Splendid, Robert. I'm proud of you."

But proud as he was, he could not leave well enough alone. He had a new idea.

"Many of your numbers after the dot have an interesting life of their own. Would you like to see what I mean?"

"Oh, yes!" said Robert. "Just so long as it won't cram the beach with famous numbers."

"Don't worry. Your big calculator will take care of them. All you need to do is enter seven divided by eleven."

Robert didn't need to be told twice.

$$7 \div 11 = 0{,}6363636363636363636 \cdots$$

"Hey, what's up? All those sixty-threes. I bet it goes on forever."

"You bet right. But that's nothing. Try dividing six by seven."

So Robert entered:

$$6 \div 7 = 0{,}857142857142857 \cdots$$

"The same numbers keep coming back!" he cried. "857142 over and over. It's, like, spinning in a circle!"

"They really are fantastic creatures, numbers! In fact, there's no such thing as an ordinary number. Every one has its own features, its own secrets. You never completely understand what makes them tick. That chain of nines after the zero, for instance: all those nines and still not quite one. And there are many others that behave even worse, that go off the deep end after their zero. They are called the unreasonable numbers, and the reason they're called that is that they refuse to play by the rules. If you have another moment for me, I'll be glad to demonstrate."

Robert knew by now that whenever the number devil made a point of being polite he was in for something. But he was so curious he couldn't resist.

"Fine," he said. "Please do."

"You recall our hopping game, I'm sure. What we did with the two and the five and the ten. Ten times ten times ten equals a thousand, which we write

$$10^3 = 1000$$

because it's faster."

"Right. And when we hop with the two, we get

$$2, 4, 8, 16, 32$$

and so on or—as always with your little tricks—till the cows come home."

"Well then," said the number devil, "how much is two to the fourth?"

"Sixteen! Pretty good, eh?"

"Perfect, my boy! And now let's do the first hop in reverse. Hopping backward, so to speak. Only when you go backward this way, you don't really hop. We call that step 'taking the rutabaga,' as if we were pulling one of those fine root vegetables out of the ground. So what is the rutabaga of four?"

"Two."

"Right! Taking the rutabaga is the reverse of our first hop. So the rutabaga of a hundred is ten, and the rutabaga of ten thousand is a hundred. What's the rutabaga of twenty-five?"

"Twenty-five," said Robert, "is five times five, which makes its rutabaga five."

"Keep it up, Robert, and you'll be my apprentice someday. Rutabaga of thirty-six?"

"The rutabaga of thirty-six is six."

"Rutabaga of 5,929?"

"Are you crazy or something?" Robert shouted. "How do you expect me to do that one? Mr. Bockel plagues us with enough dumb problems in school. I don't need to dream about them."

"Calm down, calm down," said the number devil. "Little problems like that are what the pocket calculator was made for."

"*Pocket* calculator! The thing's as big as a couch."

"Be that as it may, you'll notice it has a key with this sign on it:

Which means?"

"Rutabaga!"

"Right. Now give it a try."

$$\sqrt{5929} =$$

Robert did as he was told, and immediately read the following off the backrest of the couch calculator:

77

"Fine. But now hold on to your hat and try the rutabaga of two."

Again Robert did as he was told, and got the following:

1,4142 1356237309504880 1688724···

"Drat!" he cried. "It's utter gibberish. Number stew. I can't make head or tail of it."

"Nor can anyone else, my dear boy. That's the point. The rutabaga of two is an unreasonable number."

"Is there any way of knowing how it goes on? Because I have a feeling it does."

"Right you are, but I'm afraid I can't help you there. Taking the number any farther would mean running myself, or my calculator, into the ground."

"Wild!" Robert said. "A real monster. But write it like this:

$$\sqrt{2}$$

and butter wouldn't melt in its mouth."

"Well, let's try something a little less daunting." He drew a few figures in the sand and said, "Have a look at these:

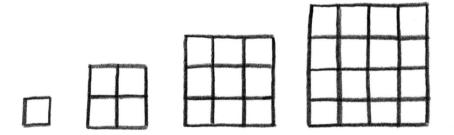

Count up the small boxes inside the squares and tell me whether you notice anything special about them."

$$1 \times 1 = 1^2 = 1$$
$$2 \times 2 = 2^2 = 4$$
$$3 \times 3 = 3^2 = 9$$
$$4 \times 4 = 4^2 = 16$$

"You bet I do. They're all hopping numbers."

"Right. You see how it works, don't you? Count the number of boxes on the side of each square and you've got the number to hop with. And vice versa. If you know how many boxes the square has—thirty-six, say—and take the number's rutabaga, you get the number of boxes along the side of the square:

$$\sqrt{1} = 1, \ \sqrt{4} = 2, \ \sqrt{9} = 3, \ \sqrt{16} = 4$$

"Great," said Robert, "but what's that got to do with unreasonable numbers?"

"Well, squares are wily beasties. Never trust a square. They may look innocent, but they can be full of tricks. Take this one, for instance." And he carved a perfectly ordinary empty square into the sand. Then he pulled a red ruler out of his pocket and laid it diagonally across it:

"Now if each side has the length one . . ."

"One what? One inch or one foot?" Robert interrupted.

"It makes no difference," said the number devil impatiently. "One whatever you please. Call it quing or quang for all I care. Now tell me how long the red line is."

"How should I know?"

"The rutabaga of two!" shouted the number devil triumphantly.

"How did you get that?" asked Robert, who was starting to feel overwhelmed again.

"Don't worry," said the number devil. "We're coming to it. All we have to do is place another square over it at an angle." He pulled five more rulers out of his pocket and laid them in the sand. Which made the figure look like this:

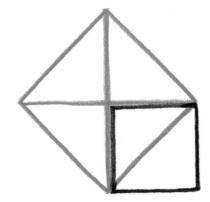

"Now guess how big the red square is, the one at an angle to the black one."

"I have no idea."

"The red square is exactly twice as big as the black one. Shift the lower half of the black one into one of the four corners of the red one and you'll see why."

"It reminds me of a game we used to play when I was little," Robert said. "Heaven and Hell we called it. We'd fold a sheet of paper into sections painted black and red. Open it to black and you went to heaven; open it to red and you went to hell."

"And do you see that in this instance there is twice as much red as there is black?"

"I do."

"Good. Now, since the area of the black square is one times one quang—we agreed to call the length of each side a quang, remember?—we can write it 1^2. And if the red square is twice as large as the black one, what is *its* area?"

"Two times 1^2," said Robert. "In other words, two."

"Correct. Then how long is each side of the red square? I'll give you a hint: all it takes is a backward hop."

"I see!" said Robert, the scales falling from his eyes. "Rutabaga! You need to take the rutabaga of two!"

"Which brings us back to our cockeyed, totally unreasonable number $1.414213\cdots$"

"Stop! Stop!" said Robert quickly. "You'll drive me crazy if you keep on with that number."

"It's not so bad as all that," said the number devil. "But we don't need to work it out. Just don't go thinking that unreasonable numbers are a rarity. Quite the contrary. Take it from me, they're like sand on the beach, more common even than the other kind."

"But there's an infinite quantity of the other kind, the ordinary ones. At least that's what you've been saying. And saying and saying."

"Because it's true. Believe me! It's just that there are many, many more unreasonable ones."

"More than what? More than an infinite quantity?"

"Exactly."

"Now you're going too far," said Robert in no uncertain terms. "I refuse to believe it. More than infinite? Nothing is more than infinite. That's a lot of malarkey, that's what it is."

"Want me to prove it?" asked the number devil. "Want me to conjure up all the unreasonable numbers at once?"

"No, anything but that! The nine chain was bad enough. Besides, what kind of proof is magic?"

"Blast!" the number devil said. "You've got me there!" But he didn't seem terribly annoyed. He merely frowned and started thinking hard.

"I could probably come up with another proof," he said at last, "but only if you insist."

"No, thank you," said Robert. "I've had enough for today. I'm beat. If I don't get a good's night sleep, I'll be in for it tomorrow in school. I think I'll stretch out on your calculator, if you don't mind. It looks awfully inviting."

"Be my guest," said the number devil as Robert lay down on the fleecy, furry, couch-sized calculator. "You're asleep as it is. You learn best when you sleep." And he tiptoed off so as not to awaken him.

"*I've had enough for today,*" *said Robert.* "*I'm beat.*" *And he lay down on the fleecy, furry, couch-sized calculator.*

Maybe he's not so bad after all, Robert thought. In fact, he's pretty cool.

Robert slept peacefully, and dreamlessly, late into the morning. He'd completely forgotten the next day was Saturday, and on Saturday, of course, there's no school.

The Fifth Night

Suddenly it was over. Robert waited and waited for his visitor from the realm of numbers, but the number devil never came. Robert went to bed as usual and most nights he had dreams, but they weren't about couch-sized calculators and hopping numbers. One was about some deep, dark holes he kept stumbling into, another about a junk room full of old suitcases with larger-than-life red ants pouring out of them: the door was locked and he couldn't get out and the ants kept crawling up his legs. In yet another he needed to ford a raging stream, but because there was no bridge he had to jump from rock to rock and just as he was about to reach the other shore all the rocks vanished and he couldn't go forward or back. Nightmares, nothing but nightmares, and no number devil in sight.

At all other times I can decide what I want to think about, Robert said to himself. Only when I dream do I have to take what comes. Why, I wonder?

"You know what?" he said one evening to his mother. "I've come to a decision. From now on—no more dreams for me."

"Glad to hear it, darling," she said. "You have trouble concentrating after a bad night, and then you don't do well in school."

Of course that wasn't what bothered Robert in the least, but all he said was "Good night." He knew you can't tell mothers everything.

But no sooner did he close his eyes than it all started up again. He was wandering through a desert where there was no shade or water, wearing nothing but a pair of bathing trunks. The heat was terrible. It was just the kind of thing he hated about dreams. On and on he trudged, thirsty, sweaty, with blisters all over his feet, until at last he made out a few trees in the distance.

It must be either a mirage or an oasis, he thought.

So on he hobbled until he came to the first palm. And there he heard a voice calling, a voice he was sure he knew.

"Hello there, Robert!" it called.

He looked up. He was right! It was the number devil, bobbing up and down on the palm leaves. He looked quite at home in the desert.

On Robert hobbled until he came to the first palm. And there he heard a voice calling, a voice he was sure he knew. He looked up. He was right!

"I'm dying of thirst," Robert called back.

"Come on up," said the number devil.

Marshaling all the strength he had left, Robert climbed to the top of the tree. His friend greeted him with a coconut and made a hole in the shell with his pocketknife.

The coconut milk tasted delicious.

"Long time no see," said Robert. "Where have you been?"

"Taking it easy, as you may have noticed."

"Any plans for today?"

"Don't you need to rest after your trek across the desert?"

"Oh, don't worry about me. I'm all right. What's the matter? No ideas?"

"I'm never at a loss for ideas, my boy."

"Something to do with numbers, I bet. It's always numbers."

"What else? There's nothing more exciting than numbers. Now what I want you to do is throw your coconut to the ground."

"Anywhere in particular?"

"No, just down."

Robert threw the coconut into the sand. From up in the tree it looked like a dot.

"Now another," the number devil ordered. "And another. And another."

"Hey, what's up?"

"You'll see soon enough."

Robert picked three coconuts and threw them all to the ground. This is what he saw in the sand:

"Keep it up," said the number devil.
So Robert threw and threw and threw.
"What do you see now?"
"Triangles," said Robert.

"Need any help?" the number devil asked, and the two of them picked and threw, picked and threw until things below looked even more triangular:

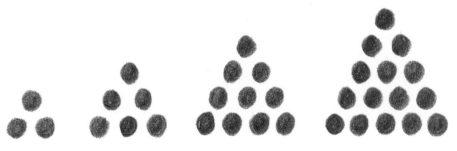

"Funny, they fall into such neat patterns," Robert said. "And I'm not even aiming. I'd never have been able to get them to land like that if I'd tried."

"Naturally," said the number devil. "You can only be that accurate in a dream—and in mathematics. Anything in mathematics can make up a nice clean pattern if you put your mind to it. By the way, we didn't really need coconuts. Tennis balls would have done just fine. Or buttons. Or chocolate-covered cherries. But now, would you count up the number of coconuts there are in each of the triangles?"

"All right. But the first is no triangle at all. It's just a dot."

"Or a triangle that has shrunk so small that all you can see of it is a dot."

"We're back to our friend one again, I see," said Robert.

"Next?"

"The second triangle has three coconuts, the third six, the fourth ten, and the fifth . . . I'm not sure. Wait. Let me count."

"Why? You don't need to count. You can calculate it."

"No I can't."

"Yes you can. Look, the first triangle, which is so small it doesn't really count as a triangle, consists of one coconut. The second has two more—the second row—which comes to:

$$1 + 2 = 3$$

The third has exactly three more—the third row:

$$3 + 3 = 6$$

The fourth has another row with four more:

$$6 + 4 = 10$$

So how many has the fifth one got?"

By this time Robert had no trouble calling it out:

$$10 + 5 = 15$$

"No need to throw down any more coconuts," he said. "I see how it works. The next triangle would have twenty-one coconuts: the fifteen from triangle number five, plus the six new ones."

"Good," said the number devil. "Now we can get down from the tree and make ourselves comfortable."

The climb down was surprisingly easy, and what did they find waiting for them on the ground but two blue-and-white-striped deck chairs and two glasses of ice-cold orange juice next to a huge swimming pool.

I can see why the number devil picked out this oasis, Robert thought. It's the ideal place for a rest.

"Now we can forget about the coconuts and concentrate on the numbers," said the number devil when both glasses were empty. "And very special numbers they are too. They're called triangle numbers, and there are more of them than you think."

"I had a feeling there'd be a lot of them," said Robert. "You do like your numbers to go on and on."

"Well, this time let's stick to the first ten. Here, let me write them down."

He got out of the deck chair, picked up his walking stick, leaned over the edge of the swimming pool, and wrote the following in the water:

1 3 6 10 15 21 28 36 45 55 . . .

Sky, sand, water—he doesn't care where he writes, Robert thought, as long as it's numbers he's writing.

"You wouldn't believe the kinds of things these triangle numbers can do," the number devil whispered in his ear. "Think of the differences, for instance."

"The differences between what?"

"Between triangle numbers next to one another."

Robert stared at the numbers floating in the pool and tried to imagine what the number devil had in mind.

1 3 6 10 15 21 28 36 45 55 . . .

"Three minus one is two. Six minus three is three. Ten minus six is four. Fifteen minus ten is five. It's just like counting from one to ten! Cool! I bet it goes on like that too."

"Precisely," said the number devil, leaning back contentedly. "But that's not all! Pick a number, any number, and I can show it to be the sum of either two or three triangle numbers."

"All right, then," said Robert. "Fifty-one."

"Simple! All I need is two—"

$$51 = 15 + 36$$

"Eighty-three."

"My pleasure—"

$$83 = 10 + 28 + 45$$

"Twelve."

"A cinch—"

$$12 = 1 + 1 + 10$$

"See what I mean? It *always* works. And now for something truly sensational. Add each triangle number to the one next to it and you won't believe what you get."

Robert stared down at the pool of floating numbers:

$$1 \quad 3 \quad 6 \quad 10 \quad 15 \quad 21 \quad 28 \quad 36 \quad 45 \quad 55 \ldots$$

And then he started adding:

$$1 + 3 = 4$$
$$3 + 6 = 9$$
$$6 + 10 = 16$$
$$10 + 15 = 25$$

"Why, they're all hopping numbers: 2^2, 3^2, 4^2, 5^2."

"Pretty good, eh? And you can go on and on."

"I believe you," said Robert. "I'd rather have a swim."

"Fine, only first let me show another one of my tricks."

"But I'm hot," Robert grumbled.

"All right then. I'll be on my way."

Now he's hurt again, thought Robert, and if I let him go, I'll probably end up dreaming about those red ants again. So he said, "No, no! Do show it to me!"

"Aha! You're curious."

"Yes, yes."

"Then pay close attention. What do you get if you add up all the ordinary numbers from one to twelve?"

"Hey, that's not like you at all! That's the kind of thing Mr. Bockel would ask." Robert glanced over at the number devil with alarm.

"Have no fear. It's simple as pie with triangle numbers. Just go to the twelfth number and you've got the sum of the numbers from one to twelve."

Robert looked down at the water again and counted:

1 3 6 10 15 21 28 36 45 55 66 78 . . .

"Seventy-eight," he said.
"Right."
"How come?"
The number devil picked up his stick and wrote:

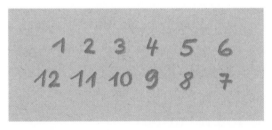

$$1 \quad 2 \quad 3 \quad 4 \quad 5 \quad 6$$
$$12 \quad 11 \quad 10 \quad 9 \quad 8 \quad 7$$

"All I've done is write the numbers from one to twelve in two rows, the first six from left to right and the second six from right to left. Draw a line under them and add them up.

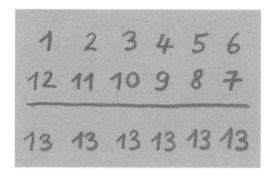

What do you get?"

"Six times thirteen," said Robert.

"I hope you don't need a calculator for that."

"Six times thirteen," said Robert, "is seventy-eight. The twelfth triangle number!"

"See what triangle numbers are good for? And by the way, quadrangle numbers aren't bad either."

"I thought we were going for a swim."

"We can swim later. First the quadrangle numbers."

Robert glanced longingly at the swimming pool, where the triangle numbers were bobbing up and down like ducklings.

"If you keep on like this, I'll wake up and all the numbers will disappear."

"The numbers *and* the pool," said the number devil. "But you can't just stop dreaming whenever you please. And who's boss here, anyway? You or me?"

There he goes again, thought Robert. Now he's losing his temper. Soon he may even start yelling at me. Sure it's all a dream, but I never like being yelled at, not even when I'm asleep. Heaven knows what he's got up his sleeve this time.

The number devil took some ice cubes from a cooler and set them out on a table in five perfect squares, each one larger than the other. "It won't take long," he said to make Robert feel better. "It's the same as with the coconuts, only this time we're using squares instead of triangles."

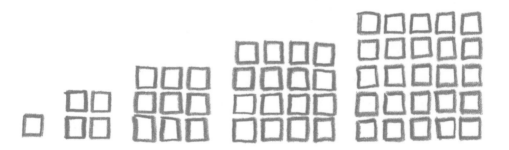

"You don't have to say another word. It's perfectly clear. They're hopped, every one of them. All I have to do is count the number of cubes on one side and hop the result:

$$1 \times 1 = 1^2 = 1$$
$$2 \times 2 = 2^2 = 4$$
$$3 \times 3 = 3^2 = 9$$
$$4 \times 4 = 4^2 = 16$$
$$5 \times 5 = 5^2 = 25$$

And so on, as usual."

"Very good," said the number devil. "Devilishly good. No, I've got to hand it to you. You're a top-notch apprentice."

> *If you're not as hot as Robert, you might want to play with the cubes a bit before they melt. Just divide up the square like this:*
>
>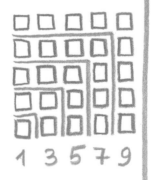
>
> *The numbers indicate the number of cubes within the lines you have drawn. What do you get if you add them together? The answer will look quite familiar.*

"Now can I have my swim?" Robert moaned.

"But surely you want to learn how the pentagonal numbers work. And the hexagonals."

"Oh, no, thank you," said Robert. "Really!" And he stood up and dived into the water.

"Wait, wait!" the number devil called out. "The pool is full of numbers. Wait until I fish them out."

But by then Robert was swimming his way through them and they were bobbing all around him, and he swam and he swam until he couldn't hear what the number devil was saying. It was a pool that went on and on just like the numbers, and it was just as exciting.

The Sixth Night

"You probably think I'm the only one," said the number devil the next time he turned up, perched on a folding chair in the middle of a vast potato field.

"The only what?" asked Robert.

"The only number devil. But I'm not. I'm one of many. Number Heaven, where I come from, is teeming with us. I'm not even one of the bosses. The bosses do nothing but sit and think. Now and then one of them will laugh and say something like 'R_n equals h_n factorial times f of n open bracket a plus *theta* close bracket,' and the others nod and laugh along. There are times when I don't understand a thing."

"You poor devil," Robert said. "Here I thought you were so sure of yourself."

"Why do you think they send *me* out at night? Because the bigwigs have things to do other than visiting apprentices like you."

"So I'm lucky to have even you. Is that what you're saying?"

"Don't get me wrong," said Robert's friend (because they were pretty much old friends by now). "I have nothing against what they cook up, the bosses up there in Number Heaven. One I particularly like is a fellow named Bonacci, an Italian, who sometimes lets me in on what he's doing. He's been dead for years now, poor Bonacci, but that doesn't matter when you're a number devil. Besides, he's a fine chap, and was one of the first to understand what zero means. He didn't discover it, mind you, but he did come up with what we call Bonacci numbers. A capital idea! And like most good ideas, it begins with—what do you think?— a one. Or, rather, two ones: $1 + 1 = 2$.

You take the last two numbers and add them together,

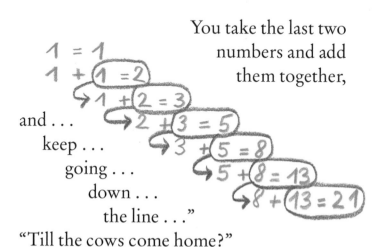

and . . .
keep . . .
going . . .
down . . .
the line . . ."

"Till the cows come home?"

"You guessed it."

Next the number devil started running through the Bonacci numbers in a kind of singsong. The aria from a Bonacci opera, you might say.

"Oneonetwothreefiveeightthirteentwentyone thirtyfourfiftyfiveeightynineonehundredandforty fourtwohundredandthirtythreethreehundredand seventyseven . . ."

Robert clapped his hands over his ears.

"All right, I'll stop," said the number devil, "though I'd better write them out so you can see what they look like."

"What have you got to write on?"

"What would you like? How about a scroll?"

Unscrewing the tip of his walking stick, he pulled out a thin roll of paper, tossed it on the ground, and gave it a poke. An endless stream of paper rolled out along a furrow.

How could all that paper fit into the stick? Robert wondered.

Meanwhile on and on it rolled until it rolled out of sight with all its Bonacci numbers:

1.	2.	3.	4.	5.	6.	7.	8.	9.	10.	11.	12.	13.
1	1	2	3	5	8	13	21	34	55	89	144	233

After that the numbers were so far off and tiny that Robert couldn't read them.

"Now what?"

"If you take the sum of the first five and add one, you get the seventh. If you take the sum of the first six and add one, you get the eighth. And so on and so forth."

"I see," said Robert, but he didn't sound particularly excited.

"It also works if you jump over numbers. Keep in mind though: the first one must always be present.

You start like this: $1 + 1 = 2$

(and now you jump one) $+ \ 3$

(and now you jump another) $+ \ 8$

(and now you jump yet another) $+21$

And what do you get when you add them up?"

"Thirty-four," said Robert.

"In other words, the Bonacci number after twenty-one. And if that's too hard for you, you can get there by hopping. For example, you take Bonacci number four, which is three, and you make it hop: 3^2. Which is . . ."

"Nine," said Robert.

"Then you take the next Bonacci number, number five, which is five, and make *it* hop."

"$5^2 = 25$," said Robert without missing a beat.

"Good, and now add the two together."

$$9 + 25 = 34$$

"Another Bonacci number!" he cried.

"And not only that," the number devil said, rubbing his hands. "The ninth. Because four and five make nine."

"Fine, fine. Fine and dandy. But tell me, what are they good for, your Bonacci numbers?"

"You don't think mathematics is for mathematicians only? Nature needs numbers too. Trees add. Fish subtract."

"Come on," said Robert. "You don't expect me to believe that."

"I expect you to believe that every living thing uses numbers. Or at least behaves as if it did. And some may well have an understanding of how they work."

"Well, I don't believe it."

"Take rabbits, for instance. They're more lively than fish. I bet there are rabbits all over this potato field."

"I don't see any," said Robert.

"Look, there are two now!"

Sure enough, two teensy white rabbits hopped up to Robert and plonked themselves at his feet.

"A male and a female, I think," said the number devil. "And a male and a female makes *one* couple. As you know, *one* is all we need to start things rolling."

"He wants me to believe you can do arithmetic," Robert said to the rabbits. "Well, I'm too smart for that."

"What do you know about rabbits, Robert?" said the two rabbits with one voice. "I bet you think we're snow rabbits."

"Snow rabbits?" said Robert, who wanted to show them that he did know something about rabbits. "Snow rabbits are winter animals, aren't they?"

"Correct," they replied, "and they're always white, while we're white only when we're young. It takes us a month to grow up, and then our fur turns brown. And then we want to have babies. It takes another month for them to be born. One boy bunny and one girl bunny."

"Just two?" asked Robert. "I always thought rabbits had oodles of bunnies."

"We have, we have," said the rabbits, "but not all at once. Two a month is enough. And they grow up and do the same. You'll see."

"But I'll have long since woken up before then. I have to go to school tomorrow morning . . . "

"No problem," the number devil interrupted. "Time runs faster here in the potato field. A month lasts only five minutes. At least when you use the special rabbit clock I just happen to have with me."

With these words, he pulled a large pocket watch out of his trouser pocket. It had two large rabbit ears, but only one hand.

"The hand shows months, not hours," he said, "and a bell rings every time a month goes past. All

I have to do to set it in motion is to press the button on top. Shall I?"

"Oh, do!" the rabbits cried.

"Good."

The number devil pressed the button. The clock started ticking; the hand started moving. When it reached one, the bell rang. A month had passed, the rabbits were much bigger and their fur had changed color—from white to brown.

When the hand reached two, two months had passed, and the mother rabbit had brought two teensy-weensy white rabbits into the world.

Now there were two couples, one younger, one older. But they did not remain satisfied for long. They wanted more babies, and by the time the hand had reached three and the bell rang again,

mother rabbit had given birth to the next two rabbits.

Robert counted the couples. Now there were three: the original one (brown), the children from their first litter, who had meanwhile grown up (and turned brown), and the furry white babies.

When the hand reached four, the old mother

rabbit gave birth to two more rabbits, and her first daughter, not to be outdone, gave birth to two. That meant that there were now five couples hopping around the potato field, three of which were brown and two of which were white.

"I wouldn't try to keep them straight if I were

you," said the number devil. "You're going to have a hard enough time just counting them."

Robert had no trouble when the clock reached five. There were only eight couples.

But there were thirteen by the sixth ring of the bell, and Robert thought, This is getting out of hand! But at least he could still count the number of

couples when the clock reached seven. There were exactly twenty-one.

"Any ideas?" the number devil asked Robert.

"Of course. It's obvious," Robert answered. "Bonacci numbers all the way:

The rabbit clock ran on and on. "Help!" Robert shouted. "Thousands of rabbits and no end in sight! This is no joke, it's a nightmare!"

1, 1, 2, 3, 5, 8, 13, 21 . . .

But even as he spoke, new hordes of rabbits were being born and joining their brown and white kin romping over the potato field. Before long Robert couldn't keep up with them anymore. And the rabbit clock ran on and on.

"Help!" Robert shouted when the hand started in on its second round. "Thousands of rabbits and no end in sight! It's awful!"

"Let me show you the rabbit list I've put together. Then you can see the whole picture. It shows everything that goes on between zero and seven."

"But it's long past seven," Robert said. "There must be thousands of them."

"4,181 couples. Which means that in five minutes there will be 6,765."

"Do you plan to let them go on? Because if you do, the whole earth will soon be covered with rabbits."

"Sooner than you think," said the number devil without batting an eyelid. "All it will take is a few rounds of the clock."

"Well, stop them! Please!" Robert begged. "This is no joke, it's a nightmare! Look, I've got nothing against rabbits. I *like* rabbits. But enough is enough. Stop them! Please!"

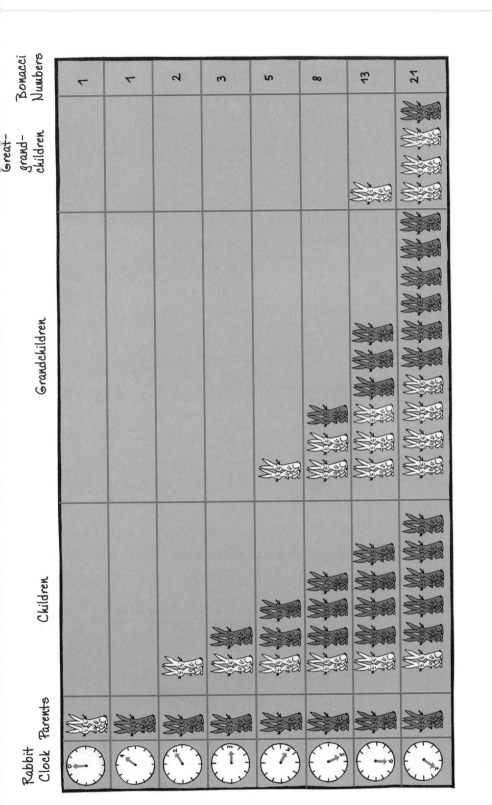

"All right, Robert. But only if you admit that the rabbits are behaving as if they had learned their Bonacci numbers by heart."

"Fine. Great. Anything you say. I admit it. Just stop them or they'll be crawling all over us in no time!"

The number devil pressed the button on the top of the rabbit clock two times, and the clock started running backward. Each time the bell rang a large number of rabbits vanished and after a few turns of the hand there were only two rabbits left in the potato field.

"What about these?" asked the number devil. "Do you want to keep them?"

"I don't think so. They'd just start over again."

"Yes, that's nature!" said the number devil, rocking gleefully on his folding chair.

"And that's how Bonacci numbers are," Robert responded. "I don't know if I like the way they take off for infinity."

"Though, as you've seen, they can just as easily go the other way. We're back where we started from. With our one."

And so again they parted in peace, leaving the rabbit couple to their own devices. The number devil went back his old friend Bonacci and their number-crunching cronies in Number Heaven.

Robert slept dreamless through the night. When the alarm rang the next morning, he was relieved to see it came from a perfectly ordinary alarm clock and not a rabbit clock.

If you still don't believe that nature acts as if it knew how numbers work, turn to the tree on the next page. Maybe you found all those rabbits a bit confusing. Well, a tree can't go romping through a field, so you won't have any trouble counting its branches. Start from below, at the red line numbered one. It runs only through the trunk, as does line two. One line higher, at line three, the trunk has been joined by a branch. Keep going. How many branches are there by the time you reach the top, line nine?

The Seventh Night

"I'm terribly worried," said Robert's mother. "I don't know what's wrong with the boy. He used to spend all his time in the park playing ball with Al and Charlie and Enrique. Now he shuts himself up in his room and spends all his time painting rabbits, rabbits, and more rabbits."

"Quiet, Mother. Please!" Robert said. "I can't concentrate."

"And the numbers he keeps muttering to himself. Numbers, numbers, and more numbers. It's not normal."

She was talking out loud to herself, as if Robert weren't in the room.

"He didn't used to be interested in numbers. You should have heard him go on about his teacher and the problems he gave him to do." Finally she turned to Robert and said, "Isn't it time you got some fresh air?"

Robert looked up from his painting and said,

"You're right. If I keep counting rabbits I'm going to get a headache."

So off he went to the park, a large grassy place with not a rabbit in sight.

"Hi!" Al called out when he caught sight of Robert. "Want to play?"

Enrique, Gary, Hugh, and Jamil were there too. They were playing football, but Robert didn't feel like joining in. They have no idea how trees grow, he thought.

It was time to eat when he got home, and he went straight to bed after supper, slipping a thick-tipped felt pen into his pajama pocket, just in case.

"Since when do you go to bed so early?" his mother asked. "You always used stay up till all hours of the night."

But Robert knew exactly what he had in mind and had no intention of telling his mother about it. She'd never believe him if he told her that rabbits, trees, and even fish understand how numbers work and that he had made friends with a number devil.

And no sooner did his head hit the pillow than the number devil was on the scene.

"Today I have something extraordinary to show you," he said.

"Anything you like, just no more rabbits. They

tortured me to death all day. I couldn't keep track of the whites and the browns."

"Forget it and come with me."

He took Robert to a white house in the form of a cube. The inside was white too, even the staircase and the doors.

"There's no place to sit," Robert complained when they went into a large, bare, completely white room. "And what are those stones doing over there?"

But when he went up to the tall pile of objects in one corner and looked at them more carefully, he realized they weren't stones at all. "They seem to be large cubes of glass or plastic," he said, "with something glittering inside. Something electric."

"Electronic," said the number devil. "What do you say we build a pyramid?" He took a few cubes and laid them out in a row along the white floor. "Well, what are you waiting for?"

Working together, they laid the following row:

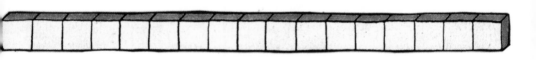

"Stop!" the number devil called out suddenly. "How many cubes have we got now?"

Robert counted them up.

"Seventeen," he said, "an unexciting number."

"More exciting than you imagine. Subtract one, for instance."

"And you get sixteen. A hopping number. A two that's been made to hop four times: 2^4."

"Good for you," said the number devil. "You're getting very observant. But let's get back to work. Each cube in the next row goes on a crack between two cubes in the first row, the way bricklayers build a wall."

"Okay," said Robert, "but it'll never be a pyramid. Pyramids are triangular or rectangular at the base, and this thing is flat. It won't be a pyramid, but it can be a triangle."

"Fine," said the number devil. "Then we'll build a triangle." Which is what they did.

"They seem to be large cubes of glass or plastic," Robert said, *"with something glittering inside. Something electric."*

"Finished!" Robert cried.

"Finished? How can that be? We're just getting started."

The number devil then climbed up one side of the triangle and wrote the number one on the top cube.

"You and your ones," Robert muttered.

"Right," the number devil replied spiritedly. "Because in the end everything always goes back to one."

"But where do we go from there?"

"You'll see, you'll see. On each cube we'll write the sum of the cubes directly above it."

"Nothing to it," said Robert, and pulling out his trusty felt pen he wrote:

"All ones," he said. "No need for the calculator yet."

"Not *just* yet," said the number devil. "Proceed."

And Robert wrote:

"Child's play," he said.

"Don't be so cocky, my boy. You've only just begun."

And Robert wrote:

"I can see that the numbers along the sides will all be ones no matter how far down we go. And that I can fill in the numbers in the next diagonal rows on either side without doing the arithmetic because they'll just be the perfectly normal numbers: 1, 2, 3, 4, 5, 6, 7 . . ."

He climbed up and down the triangle writing:

"What about the next diagonal row, the one right next to the 1, 2, 3, 4, 5, 6, 7 . . . ? Read out the first four numbers."

A knowing smile came over the number devil's face as Robert read down the row from right to left.

"1, 3, 6, 10 . . . Hey, they look familiar."

"Coconuts!" cried the number devil.

"Right, right! Now I remember. One, three, six, ten—the triangle numbers!"

"And how do you make them?"

"Sorry. That I don't remember."

"Simple."

$$1 + 2 = 3$$
$$3 + 3 = 6$$
$$6 + 4 = 10$$
$$10 + 5 = 15$$

"Fifteen and six," Robert went on, "is twenty-one."

"So you do remember!"

As a result, Robert could fill in more and more numbers. On the one hand, things got easier, because he lowered himself closer and closer to the ground; on the other hand, the numbers got awfully long awfully fast.

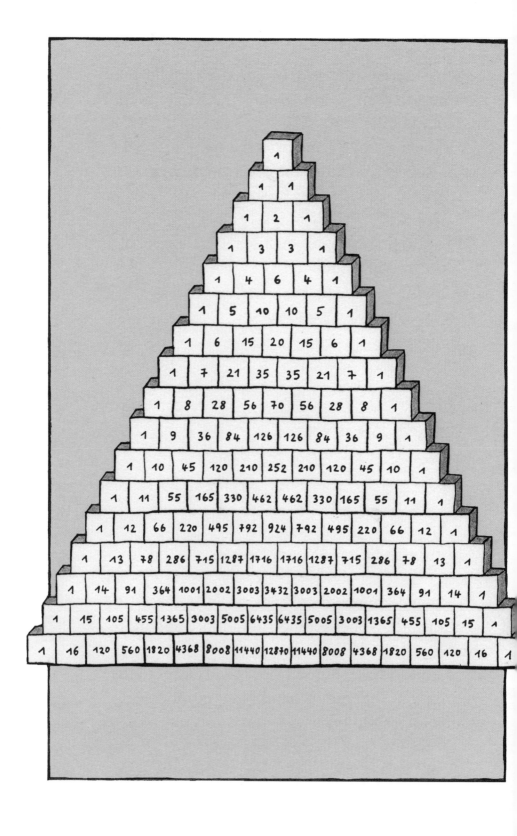

"Hey, you can't expect me to do that kind of addition in my head."

"Don't get all worked up now," said the number devil. "I wouldn't be much of a number devil if I couldn't take care of this in a flash."

And in a flash he had filled in the entire triangle.

"You really had to squeeze in that 12,870!" said Robert.

You can say that again! You may think this is only good for giving you a headache. Wrong! Quite the contrary, in fact. It's good for lazybones who don't want to bother with long sums. Let's say you need to find the sum of the first twelve triangle numbers. All you have to do is run your finger down the third diagonal row—the one that goes 1, 3, 6, 10, and so on—until you come to the twelfth cube. Then find the number just below it and toward the center. What is it?

By so doing, you have saved yourself the effort of working out what 1 + 3 + 6 + 10 + 15 + 21 + 28 + 36 + 45 + 55 + 66 + 78 comes to.

"Oh, that's nothing. There's lots more to the triangle than that! Have you any idea what we've built?" the number devil then asked. "It's more than a triangle. It's a monitor, a screen. Why do you think all the cubes have electronic insides? All I have to do is turn it on and it will light up."

With one clap of the hands he turned out the lights and with another he lit the cube on top, lit it bright red like a traffic light.

"There's that one again," said Robert.

At the next clap of his hands the first line went out and the second line glowed red.

"Would you mind adding them up for me?" the number devil asked.

"$1 + 1 = 2$," Robert mumbled. "Big deal!"

The number devil clapped his hands again, and now the third line shone red.

"$1 + 2 + 1 = 4$," said Robert. "I get it, I get it. You can stop clapping. It's our old friends, the hopping twos. The next line will be $2 \times 2 \times 2$, or 2^3, in other words, eight. And so on down the line: sixteen, thirty-two, sixty-four. Until we come to the bottom of the triangle."

"The last line is 2^{16}," said the number devil, "and that's quite a hefty number: 65,536 in case you're interested."

"I could do without it."

"Fine, fine," the number devil said, clapping his hands, and all of a sudden it was dark again. "Are you up to a visit from some more old friends?" he asked.

"It depends on who they are."

The number devil clapped three times, and the cubes instantly lit up again, though this time some were orange, others blue, and yet others green or red.

"Looks like a Christmas tree," said Robert.

"Do you see the color-coded stairs leading from the top right to the bottom left? What do you think will happen if we add up each color? Start with the red one on top."

"Which is all by itself," said Robert. "One, as always."

"Now the yellow one just below it."

"All by itself. One."

"Now we come to the blue. Two cubes."

"$1 + 1 = 2$."

"Then the green just below it. Two green cubes."

"$2 + 1 = 3$."

By now Robert knew what to expect.

"Red again: $1 + 3 + 1 = 5$. Then yellow: $3 + 4 + 1 = 8$. And blue: $1 + 6 + 5 + 1 = 13$."

"Tell me, what's going on here with this 1, 1, 2, 3, 8, 13 . . . ?"

"We're back to Bonacci and the rabbit numbers."

"See how much we've packed into our triangle? And we could go on for days. But I have a feeling you've had enough."

"More than enough, actually."

"All right then," he said, and with a clap of the hands he turned off the colored cubes. "A pity, though, because you know what I can do with just one more clap? I can light up the even numbers and leave the odd numbers dark. Are you game?"

"If you insist."

But Robert was amazed at what he saw.

"Hey, that's wild! Triangle after triangle within the triangle! Except they're upside down."

"And come in small, medium, and large. The small looks like a single cube, but it's actually a triangle; the medium consists of six cubes; the large of twenty-eight. Triangle numbers all.

"What do you think will happen if we turn off the even numbers, the numbers that can be divided by two, and light up the numbers that can be divided by three or five? All it takes is a clap of the hands. Which would you like to see? Shall we try five?"

"Yes," said Robert. "All numbers divisible by five."

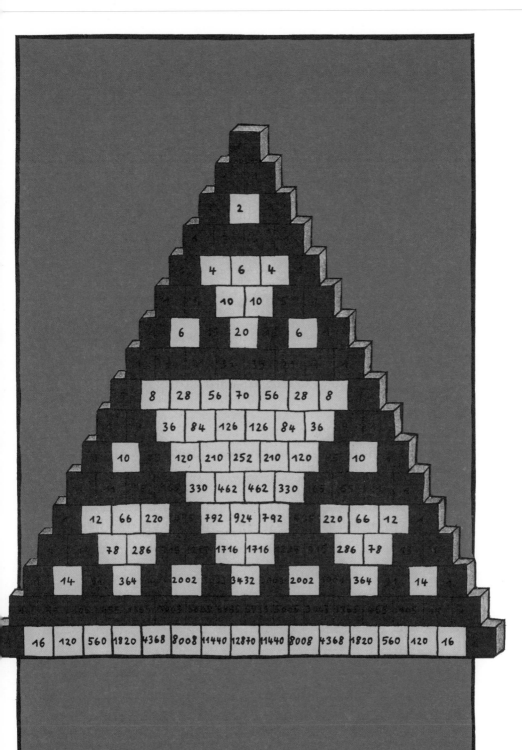

When the number devil clapped his hands, the orange lights went out and green lights came on.

"Never in my wildest dreams would I have expected triangles again," said Robert. "The same, but different. Pure witchcraft!"

"Yes, my boy. I often wonder where mathematics stops and witchcraft begins."

"Fantastic! Is this all your doing?"

"No."

"Well then, whose?"

"The devil only knows. The great number triangle goes back a long way. It's much older than I am."

"And you're no spring chicken yourself."

"Me? How can you say such a thing! Why, I'm one of the youngest residents of Number Heaven. And our triangle is at least two thousand years old. It was a Chinese gentleman who came up with the idea, I believe. But we still enjoy playing with it and making it do new tricks."

Nothing you do ever seems to have an end to it, thought Robert, not daring to say it out loud.

But the number devil must have read his thoughts, because he said, "Yes, mathematics is an endless story. Keep digging and you keep coming up with new things."

"You mean you can't stop?" Robert asked.

"I can't," whispered the number devil, "but you . . ." And as he spoke, the green cubes grew paler and paler and he grew thinner and thinner until they went off altogether and he was only a shadow of his former self. Before long, Robert had forgotten everything: bright cubes, big triangle, Bonacci numbers, and even his friend the number devil.

"You're looking pale this morning," his mother said when he awoke after a long, long sleep. "Are you having nightmares again?"

"Not in the least."

"Well, I'm worried about you."

"Don't be, Mother. You know the saying: The Devil is never so grim as he is painted."

Are any of you curious about what kind of pattern we get when we light up all the numbers that can be divided by four? You don't need to be a number devil to figure it out. Just copy the triangle on the next page, and take a colored pencil and color in all the numbers that occur in the fours column of the multiplication table. With numbers above forty-eight use your calculator: enter the number, then ÷, 4, and =, and see whether it comes out even.

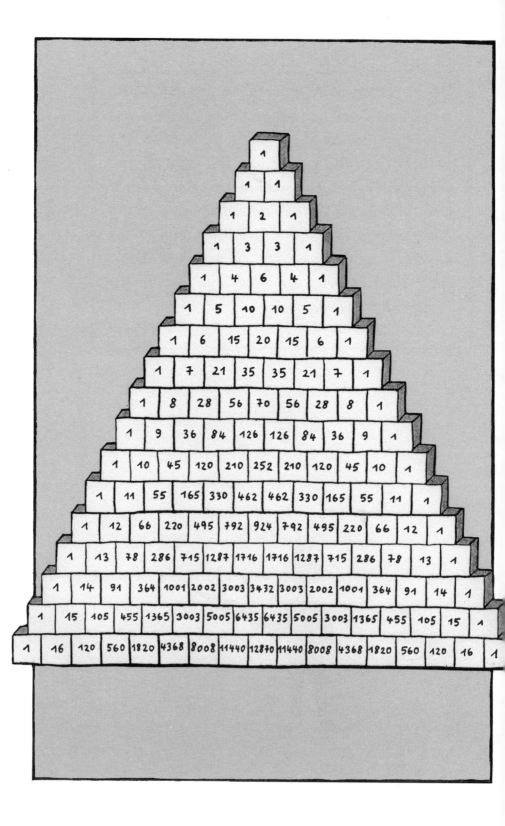

The Eighth Night

Robert was up at the blackboard. His two best friends in the class—Al, the one he played football with, and Betsy, the one with the pigtails—were sitting in front, as usual. And, as usual, they were having an argument.

Just what I needed, thought Robert. A dream about school!

At that very moment the door flew open and in came not Mr. Bockel but—the number devil.

"Good morning," he said. "Arguing again, I see."

"Betsy's sitting in my place," said Al.

"Then switch places."

"She won't budge."

"Put an *A* for Al and a *B* for Betsy on the board, Robert," said the number devil.

Why not, Robert thought, if it makes him happy.

"Now, Betsy," said the number devil, "I want you to change places with Al."

And for some strange reason Betsy didn't make a scene. She stood up and changed places with Al.

B A

wrote Robert on the board.

At that very moment the door flew open again and in came Charlie, late as usual. Charlie sat down next to Betsy.

C B A

wrote Robert.

But Betsy didn't like that. "If I'm going sit on the left, I'm going to sit all the way on the left."

"Heavens to Betsy!" Charlie cried.

And the two changed places:

B C A

Then Al was upset. "I want to sit next to Betsy!" he said, so easygoing Charlie gave Al his place without a murmur:

If this keeps up, Robert said to himself, we can forget about the class.

And it did keep up, because Al decided *he* wanted to sit on the left.

"Which means we all have to get up," said Betsy. "I don't know why, but . . . Come on, Charlie."

Here is how it looked when they'd all settled down again:

Not that it lasted, of course.

"I won't sit next to Charlie for another minute," said Betsy before long. She was a real pain, that Betsy. And since Betsy had to have her way, the boys had to get up and move. Robert wrote:

CAB

"Enough is enough," he said. "That's it."

"Is it, now?" the number devil responded. "The three of them haven't exhausted the possibilities open to them. Supposing Al were to sit on the left, Charlie in the middle, and Betsy on the right?"

"Not on your life!" Betsy cried.

"Don't make such a fuss, Betsy!"

The three of them pulled themselves up again and repositioned themselves:

A C B

"Hey, Robert! Notice anything? I don't think those three will come up with anything else."

Robert looked at the board:

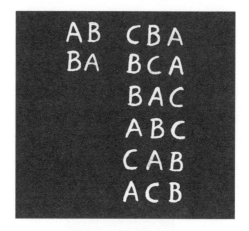

"We seem to have tried all the possibilities," he said.

"That we have," said the number devil. "But there are more than the four of you in this class. Quite a few more, I'm afraid."

At that very moment the door flew open and in ran Doris, out of breath.

"Hey, what's going on here? Where's Mr. Bockel? And who are you?"

"I'm substituting for Mr. Bockel. He's taking the day off. He said he needed a rest from the chaos that reigns in this class."

"I can see why," said Doris. "Look at them. They're all in the wrong seats. Since when do you sit *there*, Charlie? That's my desk!"

"What order do you suggest, Doris?" the number devil asked.

"Why not go by the alphabet?" she said. "*A* for Al, *B* for Betsy, *C* for Charlie, and so on. That's the simplest way out."

"As you like. Let's give it a try."

So Robert wrote the following on the board:

A B C D

But the others weren't the least bit happy with Doris's suggestion, and all hell broke loose in the classroom. Betsy was the worst, biting and scratching when one of the others refused to give way, but they all pushed and shoved. Then, little by little, the crazy game of musical chairs they were playing began to seem like fun, and they switched places so fast that Robert had trouble keeping up with them. In the end, though, he did manage to get all their seating combinations on the board:

ABCD	BACD	CABD	DABC
ABDC	BADC	CADB	DACB
ACBD	BCAD	CBAD	DBAC
ACDB	BCDA	CBDA	DBCA
ADBC	BDAC	CDAB	DCAB
ADCB	BDCA	CDBA	DCBA

It's a good thing there are lots of kids absent today, thought Robert, or it would go on forever.

And at that very moment the door flew open and in came Enrique, Felice, Gary, Hugh, Iris, Jamil, and Karen.

"No! No!" Robert cried. "No! Please! Don't sit down or I'll go crazy!"

"All right," said the number devil. "That's it for today. You can all go home."

"Me too?" Robert asked.

"No, I need you to stay a while."

While his classmates ran out into the schoolyard, Robert looked over the numbers on the board.

"Well," said the number devil. "What do you make of them?"

"I'm not sure," said Robert. "All I know is that the number of ways they can sit increases awfully fast. As long as there were only two kids, things

"No! No!" Robert cried. "No! Please! Don't sit down or I'll go crazy!"
"All right," said the number devil. "That's it for today. You can all go home."

were simple: two kids, two possibilities. But with three kids there were six possibilities. And with four—wait a minute—twenty-four."

"What if there's only one?"

"What a thing to ask! Only one possibility."

"Let's try multiplying," said the number devil.

Children:	Possibilities:
1	1
2	$1 \times 2 = 2$
3	$1 \times 2 \times 3 = 6$
4	$1 \times 2 \times 3 \times 4 = 24$

"I see," said Robert. "Interesting."

"The more of your classmates join in the game, the more inconvenient it is to write it out like that. There's a shorter way, though. You take the number and put an exclamation mark after it. Like this:

$$4! = 24$$

And you read it: four vroom!"

"What do you think would have happened if you hadn't sent Enrique, Felice, Gary, Hugh, Iris, Jamil, and Karen home?"

"Utter confusion. Pandemonium," replied the number devil. "I can just see them pushing and shouting, trying out each and every combination. It would have taken ages. Together with Al, Betsy, and Charlie there would have been eleven of them. That means eleven vroom! possibilities. Can you guess how many that is?"

"I know I can't do it in my head, but I always bring my calculator to school—I have to hide it, of course: Mr. Bockel can't stand the sight of calculators—so I'll have the answer for you in a jiffy.

$$1 \times 2 \times 3 \times 4 \times 5 \times 6 \times 7 \times 8 \times 9 \times 10 \times 11 =$$

"Eleven vroom!" he said excitedly, "is precisely 39,916,800. Wow! Nearly 40 million!"

"So if we'd gone through all the combinations, we'd still be here eighty years from now. Your classmates would be in wheelchairs and we'd have to hire eleven nurses to do the pushing. See how useful a bit of mathematics can be? Which reminds me . . . Have a look out of the window and tell me if your classmates are still there."

"Oh, I'm sure they've all gone their separate ways."

"I assume you shake hands when you say good-bye."

"Shake hands! We mumble, 'See you'—if you're lucky."

"A pity," said the number devil, "because I wonder how long it would take for each of them to shake hands with each of the others."

"You know perfectly well there'd be an untold number of handshakes. Eleven vroom! of them, I suppose, since there are eleven of them."

"Wrong!" said the number devil.

"Wait a minute," said Robert. "I see. If there were two of them, they'd need only one handshake. If there were three . . ."

"Try putting it on the board."

This is what Robert wrote:

People:	Handshakes:
A	—
A B	A B
A B C	A B A C B C
A B C D	A B A C A D B C B D C D

"Two people—one handshake. Three people—three handshakes. Four people—six handshakes. Five people—ten."

"One, three, six, ten . . . Look familiar?"

Robert couldn't remember, so the number devil made a few big dots on the board:

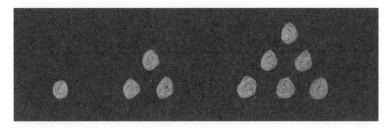

"Coconuts!" Robert shouted. "Triangle numbers!"

"And how do they go?"

Robert wrote on the board:

$$1 + 2 \ = 3$$
$$3 + 3 \ = 6$$
$$6 + 4 \ = 10$$
$$10 + 5 \ = 15$$
$$15 + 6 \ = 21$$
$$21 + 7 \ = 28$$
$$28 + 8 \ = 36$$
$$36 + 9 \ = 45$$
$$45 + 10 \ =$$

"So you'd need exactly fifty-five handshakes."

"That wouldn't be so bad," said Robert.

"And here's what you do to get around all that

arithmetic. You draw a few circles on the board:

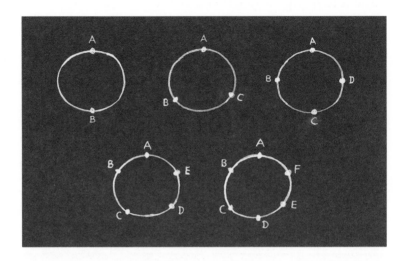

The letters stand for your friends: *A* for Al, *B* for Betsy, *C* for Charlie, and so on.

"Then you join the letters with lines:

"Pretty, isn't it? And since each line represents a handshake, all you have to do is count them."

"One, three, six, ten, fifteen . . . As usual," said Robert. "There's only one thing I don't understand. How is it that everything you do works?"

"That's the devilish thing about numbers: everything works. Well, almost everything. Because the prima-donna numbers—remember them?—they have their problems. And you've got to keep your eyes peeled for them or you'll fall flat on your face. Which is why so many people hate numbers. I can't stand slobs, and they can't stand numbers. By the way, go to the window and you'll see a schoolyard that looks like a pigsty."

Robert had to admit it. The schoolyard was strewn with soda cans, newspapers, and sandwich wrappings.

"If three of you pick up some brooms, you can sweep it clean in half an hour."

"Which three have you got in mind?" Robert asked.

"Al, Betsy, and Charlie, say. Or Doris, Enrique, and Felice. And you've got Gary, Hugh, Iris, Jamil, and Karen waiting in the wings."

"So *which* three doesn't matter."

"Right."

"Then we can combine them any old way," said Robert.

"Right again. But supposing they're not all available. Supposing Doris, Enrique, and Felice aren't
available, so we have only three: Al, Betsy, and
Charlie."

"Then they'll have to do it."

"Good. Put that on the board."

And Robert wrote:

$$ABC$$

"Now, if Doris runs in late as usual, what do we
do? What are the possibilities?"

Robert thought a moment and wrote the
following:

$$ABC \quad ABD \quad ACD \quad BCD$$

"These four," he said.

"Now Enrique turns up. Why shouldn't he be
included? That makes five candidates. See what
you can do with five."

But Robert refused. He was getting a little
nervous.

"You tell me," he said.

"All right. With three people we can have only
one group of three; with four we can have four.
And with five we can have ten. Here, let me put it
on the board:

People: Groups:

3	ABC									
4	ABC	ABD		ACD			BCD			
5	ABC	ABD	ABE	ACD	ACE	ADE	BCD	BCE	BDE	CDE

"There's something special I want you to notice. As you can see, I've put the groups in alphabetical order. How many groups begin with Al? Ten. How many with Betsy? Four. And with Charlie only one. The same numbers keep coming up:

$$1, 4, 10 \cdots$$

Can you guess how it goes on from there? I mean, if we add a few more names: Felice, Gary, Hugh, and so forth. How many groups would we have then?"

"Beats me," said Robert.

"You remember how we cracked the handshake

problem? When everybody shakes hands with everybody else?"

"That was a breeze. We used the triangle numbers:

$$1, 3, 6, 10, 15, 21 \cdots$$

But they won't help with our broom brigades, which work three to a group."

"Now what if you add the first two triangular numbers together?"

"That makes four."

"And the next one."

"That makes ten."

"And the next."

"$10 + 10 = 20$."

"Go on."

"You mean keep going until I get to the eleventh? You can't be serious."

"Don't worry. You can get there without arithmetic. You can get there without guessing, even without *ABCDEFGHIJK*."

"How?"

"With our good old number triangle," said the number devil.

"You mean you're going to put one on the board?"

"Heavens no! Not when I have my walking stick handy."

No sooner did he tap the stick on the board than there it was, in all its glory. In glorious color too.

"Couldn't be easier," he said. "For the handshakes you count from top to bottom using the green cubes: one handshake for two people, three for three people, and fifty-five for eleven people.

"For our broom-brigade trio you use the red cubes, again from top to bottom. We start out with three people and one possibility. When we have four people to choose from, we have four combinations. With five people we have ten. How many would we have if all eleven of your classmates show up?"

"165," Robert answered. "You're right. It *is* easy. This number triangle is nearly as good as a computer. But tell me, what are the orange cubes for?"

"Oh, them," said the number devil. "Well, as you may have noticed, we number devils aren't easily satisfied; we tend to go overboard. So just in case three people couldn't handle the clean-up and you needed a fourth, I wanted you to know how many possibilities you'd have. How many will there be if, say, eight people apply for a broom-brigade quartet?"

"Seventy," said Robert, having no trouble finding the answer in the number triangle.

"Correct," said the number devil. "By now you can guess what the blue cubes are for."

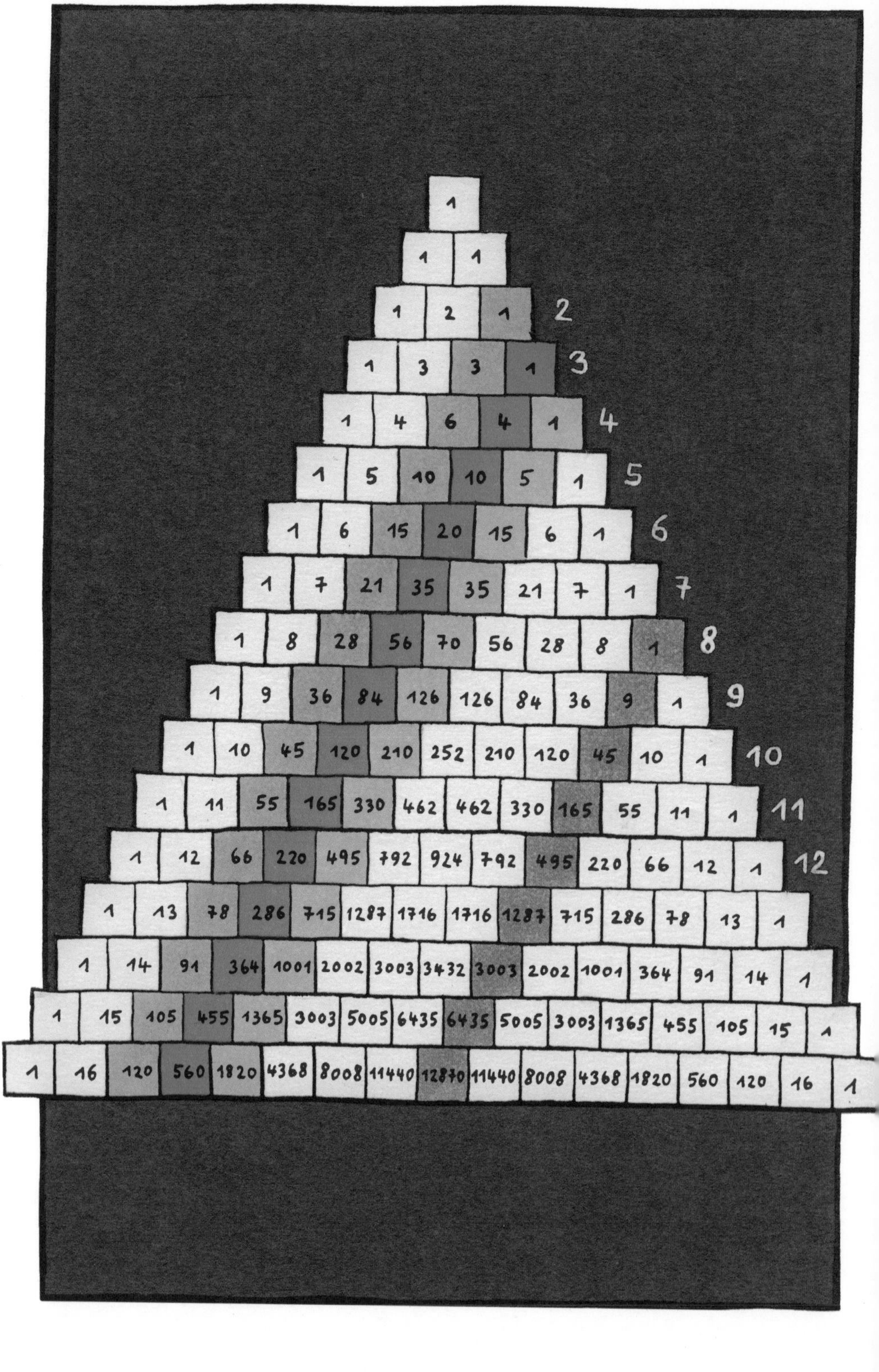

1
1 1
1 2 1 2
1 3 3 1 3
1 4 6 4 1 4
1 5 10 10 5 1 5
1 6 15 20 15 6 1 6
1 7 21 35 35 21 7 1 7
1 8 28 56 70 56 28 8 1 8
1 9 36 84 126 126 84 36 9 1 9
1 10 45 120 210 252 210 120 45 10 1 10
1 11 55 165 330 462 462 330 165 55 11 1 11
1 12 66 220 495 792 924 792 495 220 66 12 1 12
1 13 78 286 715 1287 1716 1716 1287 715 286 78 13 1
1 14 91 364 1001 2002 3003 3432 3003 2002 1001 364 91 14 1
1 15 105 455 1365 3003 5005 6435 6435 5005 3003 1365 455 105 15 1
1 16 120 560 1820 4368 8008 11440 12870 11440 8008 4368 1820 560 120 16 1

"The broom-brigade octet," said Robert. "If I have eight volunteers, I have only one possibility. But with ten I have forty-five. And so on."

"You get the picture."

"What do you think the schoolyard looks like now?" Robert said, looking out of the window. It was cleaner than he had ever seen it. "I wonder which three did it?"

"Well, you weren't among them, my dear Robert."

"How do you expect me to sweep the schoolyard when you keep throwing numbers and cubes at me all night long?"

"Maybe you need a little rest from me."

"What do you mean? Aren't you coming back?"

"I thought I could use a little holiday," said the number devil. "You can always talk numbers with Mr. Bockel."

It wasn't the greatest prospect in the world, but he had no choice. And he had to go back to school the next day anyway.

When he walked into the classroom, he saw Al and Betsy in their normal places. Neither seemed keen on switching.

"Here comes the wizard!" Charlie called out.

"Robert does problems in his sleep!" Betsy teased.

"Do you think it helps?" asked Doris.

"Not very much," said Jamil. "Mr. Bockel can't stand him."

"Well, the feeling's mutual," Robert replied.

Robert stole a glance at the schoolyard.

As usual, he thought. A regular dump! So much for his dream. But the numbers remained. He could count on the numbers.

At that moment the door flew open and in walked the inevitable Mr. Bockel with a briefcase chock-full of pretzels.

The Ninth Night

Robert dreamed he was dreaming. He was used to it by now. Whenever he had a dream about something unpleasant—the one about being stranded on a slippery rock in the middle of a raging stream, for instance—he would think to himself, Horrible as it is, it's all a dream.

But one day he caught the flu, and, lying in bed all day with a temperature, he found the trick didn't work. Besides, fever dreams were the worst. The last time he'd been sick in bed he'd had a dream about a volcano erupting. Fire-spewing mountains had flung him into the air, and he was about to descend slowly—slowly, how curious— into the maw of the volcano. It gave him the creeps just to think of it. So he tried to stay awake, even though his mother kept telling him, "The best thing to do is to sleep it off. Don't read so much. It's unhealthy!"

After the twelfth comic book, however, his eyes were so heavy that they closed by themselves, and what he dreamed was as strange as strange could be.

He dreamed he was in bed with the flu and the number devil was sitting next to him. There was a glass of water on his bedside table and he thought, I'm hot. I have a temperature. I don't think I'm asleep.

"What about me?" asked the number devil. "Are you dreaming me or am I really here?"

"I'm not sure," said Robert.

"What difference does it make?" said the number devil. "I'm just making a sick call. And since you're ill and must stay in bed and can't climb trees in the desert or count rabbits in the country, I thought we'd spend a quiet evening at home. I've brought along a few numbers to take your mind off things. They're perfectly harmless, I assure you."

"That's what you always say."

Just then there was a knock on the door.

"Come in!" the number devil called out.

And in they marched. They reminded him of racing cyclists or marathon runners, because they sported their numbers on white T-shirts. They came in such quantities that before he knew it,

Robert's room was packed. At first he was amazed to see so many squeeze into so small a space, but then he realized that as more and more crowded in, the door moved farther and farther away, until it stood at the end of a long, narrow corridor and he could hardly make it out.

For a while the numbers just stood there laughing and chattering away. Then the number devil shouted in his best army-sergeant voice, "Attention! First row, fall in!" and they immediately lined up, backs to the wall, one at the head, the others following in numerical order.

"Where's zero?" Robert asked.

"Zero, front and center!" the number devil roared.

Zero had hidden under the bed and crawled out, terribly embarrassed.

"I thought I'd not be needed. I'm not myself today. I must be coming down with the flu. I'm afraid I'll have to ask for sick leave."

"Dismissed!" the number devil shouted, and zero crept back under Robert's bed.

"That zero! Always making problems, wanting something special. But the others—I hope you appreciate how well they follow orders."

He seemed tremendously pleased with a line of perfectly ordinary numbers:

1	2	3	4	5	6	7	8	9	10	11	12	13	...

"Second row, fall in!" he shouted, and immediately a new contingent of numbers stormed in and found their places with a great clatter and shuffle:

They stood directly in front of the others, making the room look even more like an interminable tunnel. They were all decked out in identical red T-shirts.

"I see," said Robert. "The odd numbers."

"Right. Now I want you to guess how many of them there are compared with their white-shirted comrades along the wall."

"That's obvious," said Robert. "Every other number is odd, so there are half as many reds as there are whites."

"What you're saying is that there are twice as many ordinary numbers as odd."

"Right."

The number devil laughed, but it wasn't a nice laugh. Robert thought it sounded sarcastic.

"Sorry to disappoint you, my boy, but as you see, there are exactly the same number of one as of the other."

"Ridiculous!" Robert cried. "*All* can't be the same as *half*."

"Watch carefully and I'll show you what I mean."

He turned to the numbers and roared, "First and second rows, shake hands!"

"You don't need to scream at them, do you?" Robert said angrily. "This isn't an army barracks. Try being a little more polite."

But his protest went unheeded, because by then they had formed pairs like tin soldiers and each white was shaking hands with a red:

1	2	3	4	5	6	7	8	9	10	11	12	13	...
1	3	5	7	9	11	13	15	17	19	21	23	25	...

"See? Each ordinary number from one on has its own odd number from one on. Can you show me a single red without a partner? So there is an infinite quantity of ordinary numbers and an infinite quantity of odd numbers. Infinite, understand?"

Robert thought for a while.

"So if I divide an infinite quantity in half I get two infinite quantities. But then the whole is the same size as the half."

"Correct," said the number devil. "And not only that." He pulled a whistle out of his pocket and gave a toot. All at once a new column of numbers—this one in green T-shirts—appeared out of the depths of the endless room, jiggling and joggling until the number devil commanded, "Third row, fall in!"

In a flash the greens formed a neat line in front of their red and white comrades:

"Prima donnas," Robert concluded from the numbers on their T-shirts.

The number devil merely nodded. Then he gave another toot on his whistle and another and another and another. All hell broke loose. A nightmare! Who would have thought that so many numbers could fit in a single room, even if it had by now grown as long as the path a rocket takes to the moon. There was no air left. Robert's head felt like a glaring lightbulb.

"Stop! Stop! I can't take any more of this!"

"Come in!" the number devil called out, and in marched the numbers in such quantities that before Robert knew it, his room was packed.

"Your flu must be getting to you," said the number devil. "I'm sure you'll feel better tomorrow."

Then he turned to the numbers and shouted, "Now hear this! Rows four, five, six, and seven, fall in! On the double!"

Robert forced his drooping eyes open and saw seven kinds of numbers in white, red, green, blue, orange, black, and pink T-shirts standing one behind the other in neat but endless rows:

1	2	3	4	5	6	7	8	9	10	11	12	13	14	15	...
1	3	5	7	9	11	13	15	17	19	21	23	25	27	29	...
2	3	5	7	11	13	17	19	23	29	31	37	41	43	47	...
1	1	2	3	5	8	13	21	34	55	89	144	233	377	610	...
1	3	6	10	15	21	28	36	45	55	66	78	91	105	120	...
2	4	8	16	32	64	128	256	512	1024	2048	4096	8192	16384		...
1	2	6	24	120	720	5040	40320		362880		3628800	39916800			...

The numbers on the pink T-shirts were soon so long that they barely fit, and it was all Robert could do to read them.

"They get large so quickly! I'll never keep up."

"Vroom!" said the number devil. "The numbers with the exclamation mark:

$$3! = 1 \times 2 \times 3$$
$$4! = 1 \times 2 \times 3 \times 4$$

And so on. Hard to keep up with them, isn't it? But what about the others? Do you recognize them?"

"Let's see. The reds are odd, the greens prima donnas, the blues—I don't know, but they look familiar."

"Think rabbits."

"Oh, yes. The Bonaccis. Which would make the orange numbers triangle numbers."

"Not bad. Flu or no flu, you're making progress."

"The blacks are obviously hopping numbers: 2^2, 2^3, 2^4, and so on."

"And there is an equal quantity of each color," said the number devil.

"An infinite quantity," said Robert with a sigh. "Awful, isn't it. A real mob scene."

"Rows one through seven, dismissed!" the number devil roared.

In a flurry of scraping, pushing, puffing, and trampling, the numbers left the room. They were replaced by an exquisite stillness. Robert's room was as small and bare as before.

"All I need now is an aspirin and a glass of water."

"Plus a good rest, and you'll be back on your feet tomorrow," said the number devil, gently tucking Robert in.

"But do you think you can keep your eyes open long enough to take care of what we have left?"

"Left of what?"

"The thing is," he said, waving his stick again, "we booted the numbers out because they made such a mess of your room. But we still have series to deal with."

"Series? What are series?"

"Well, you don't think numbers just stand there like tin soldiers, do you? What happens when they are combined—that is, when they're added together?"

"I don't know what you're talking about," Robert moaned.

But by then the number devil had drawn the first series on the ceiling with his stick.

"I thought you said I needed rest."

"This won't take much out of you. All you have to do is read what it says."

"Fractions!" Robert moaned again. "Yuck!"

"What do you mean? What could be simpler than fractions? Look at these!"

When Robert looked up at the ceiling again, the numbers were gone and had been replaced by a long line.

$$\frac{1}{2} + \frac{1}{4} + \frac{1}{8} + \frac{1}{16} + \frac{1}{32} + \frac{1}{64} \ldots =$$

"One-half," Robert read, "plus one-quarter, plus one-eighth, plus one-sixteenth, plus one-thirty-second, and so on. Ones on top and hopping twos on the bottom. The same as the black T-shirts: two, four, eight, sixteen, thirty-two ... And I'm sure I know what comes next."

"Yes, but what comes from adding them all together?"

"I have no idea, though if the series never ends then what comes of it must be never-ending too. On the other hand, one-quarter is less than one-half, one-eighth is less than one-fourth, and so on. So the numbers I add on will get smaller and smaller."

When Robert looked up at the ceiling again, the numbers were gone and had been replaced by a long line:

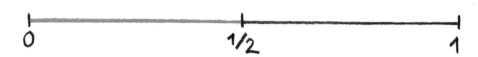

0   1/2   1

"I think I get it," he said, after staring up at it for a while. "You start with one-half, then add on half of one-half, in other words, one-quarter."

As he spoke, the numbers appeared in black and white on the ceiling:

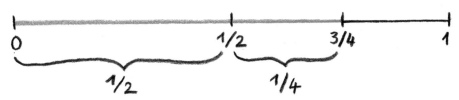

"And you just keep going, adding on half of the previous number. Half of one-quarter is one-eighth, half of one-eighth is one-sixteenth, and so on. The pieces will get smaller and smaller, so small that they'll soon be invisible, like the pieces of chewing gum you divided up that first night.

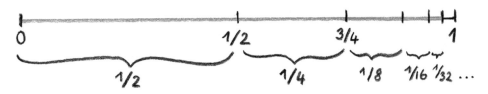

You can keep going till you're blue in the face and you'll never reach the one. Almost, but never quite."

"Well, I want you to keep going."

"And I don't want to. I'm in bed with the flu, remember?"

"That's just the point. *You* may get tired, but numbers don't," the number devil said. "They can go on and on forever."

Suddenly the line on the ceiling was replaced by the following:

$$\frac{1}{2} + \frac{1}{4} + \frac{1}{8} + \frac{1}{16} + \frac{1}{32} + \frac{1}{64} \cdots = 1$$

"Well done!" cried the number devil. "Excellent! Keep going."

"But I'm tired. I need to sleep."

"Sleep?" said the number devil. "You *are* sleeping. You're dreaming of me, aren't you? And you can only dream if you're sleeping."

There was nothing Robert could say to that, though he felt his brain was turning into jelly.

"All right, I'll go along with *one more* of your crazy ideas, but then I've got to rest."

The number devil raised his stick, snapped his fingers, and a whole new series appeared on the ceiling:

$$\frac{1}{2} + \frac{1}{3} + \frac{1}{4} + \frac{1}{5} + \frac{1}{6} + \frac{1}{7} + \frac{1}{8} + \cdots =$$

"It's just like the last one," Robert said. "I can go on adding till the cows come home, but since each number is smaller than the one before it, they'll never add up to one."

"Is that what you think? Then let's look a bit closer. At the first two numbers, for instance."

Now only the first two numbers of the series were left on the ceiling:

$$\frac{1}{2} + \frac{1}{3}$$

"What's the answer?"

"I don't know," Robert muttered.

"Don't act stupid now. Which is more? One-half or one-third?"

"One-half, of course!" Robert said, annoyed. "What do you take me for?"

"Now, now. Just tell me this: Which is more? One-third or one-fourth?"

"One-third, of course."

"So we have two fractions, both of which are more than one-fourth. And what do two-fourths make?"

"What a dumb question! Two-fourths make a half."

"Good.

$$\frac{1}{2} + \frac{1}{3}$$ is therefore more than $$\frac{1}{4} + \frac{1}{4}$$

And if we take the next four terms of the series and add them together, they too come out to be more than one-half. Look:

$$\frac{1}{4} + \frac{1}{5} + \frac{1}{6} + \frac{1}{7}$$

"That's too complicated for me," Robert grumbled.

"Nonsense!" cried the number devil. "Which is more? One-fourth or one-eighth?"

"One-fourth."

"Which is more, one-fifth or one-eighth?"

"One-fifth."

"Right. And the same holds for one-sixth and one-seventh. See these following fractions?

$$\frac{1}{4}, \; \frac{1}{5}, \; \frac{1}{6}, \; \frac{1}{7}$$

They are all more than one-eighth. And what do four-eighths make?"

Robert did not even want to answer, but he finally said, "Four-eighths make exactly one-half."

"Excellent. So now we have

$$\underbrace{\frac{1}{2}+\frac{1}{3}}_{} + \underbrace{\frac{1}{4}+\frac{1}{5}+\frac{1}{6}+\frac{1}{7}}_{} + \underbrace{\frac{1}{8}+\frac{1}{9}+\frac{1}{10}+\frac{1}{11}+\frac{1}{12}+\cdots\frac{1}{15}}_{}+\frac{1}{16}\cdots$$

more
than ½

more
than ½

more
than ½

And so on. Till the cows come home. You'll notice that if we add the first six terms of the series together they come to more than one. And we can go on like this as long as we like."

"No, no!" Robert cried. "Please!"

"But *if* we went on—don't worry, we won't— where would it take us?"

"To infinity, I suppose," said Robert. "How devilish of you."

"Except that it would take forever. Even if we worked at lightning speed, we wouldn't reach the first thousands till, say, the end of the world. That's how slowly the series increases."

"Then let's leave well enough alone."

"Yes, let's leave well enough alone."

And with that the writing on the ceiling began to fade, the number devil grew thinner and thinner, and time moved on.

Robert did not wake up until the sun was tickling his nose.

"Thank God the fever is gone," his mother said, putting her hand on his forehead.

By then he had forgotten how easy and how slow it can be to slide from one to infinity.

The Tenth Night

Robert was sitting on his backpack in the middle of the snow. He had no idea where he was, but it felt like the North Pole. The cold had crept into his hands and feet, and the snow showed no sign of letting up. It was a real blizzard! No light, no house, no living being as far as the eye could see, and night was coming on. Unless something happened mighty quickly, he was doomed.

Yet even as he tried to warm his stiff, blue hands by clapping them together—he didn't want to freeze to death, after all!—he was aware of another Robert sitting perfectly content in an armchair watching him shudder. So you can dream of yourself dreaming, Robert thought.

Then the snowflakes swirling around the face of the freezing Robert started growing in size, and the other, warm, Robert, lolling comfortably in his armchair, noticed that no snowflake was like any other. Every one of the large soft flakes was unique. Most had six sides or points, and when Robert looked closer he saw that certain patterns tended to return: hexagonal stars in a hexagonal star, points branching off into smaller and smaller points . . .

Suddenly he felt a hand on his shoulder and heard a familiar voice. "Beautiful, aren't they?" it said.

It was the number devil. He was sitting right behind him.

"Where am I?" Robert asked.

"Just a second," the number devil answered. "I'll turn the light on."

Hexagonal stars in a hexagonal star, points branching off into smaller and smaller points . . . A familiar voice said, "Beautiful, aren't they?"

All at once there was a dazzling light, and Robert saw that he was sitting in a small, elegantly appointed auditorium with only two rows of red plush seats.

"A private showing," said the number devil. "Just for you."

"And I thought for sure I was going to freeze to death."

"It was only a film. Here, I have something for you."

It was not another pocket calculator. It was not a sticky green or a furry couch calculator. No, it was a silver-gray beauty, complete with a nifty mouse and a flip-top monitor.

"A computer!"

"Just a little notebook. Only it's rigged up to project everything you input onto the screen in the front of the auditorium. Which means you can draw directly on the screen with your mouse. Shall we begin?"

"Okay, but no more snowstorms, promise? Numbers, no North Pole."

"Bonacci numbers?"

"You and your Bonacci!" exclaimed Robert. "Tell me, is that guy your best friend or something?"

The numbers flashed on the screen as he entered them:

1, 1, 2, 3, 5, 8, 13, 21, 34, 55, 89 · · ·

"Now try dividing them by their neighbors," the number devil suggested. "The larger by the smaller."

"Okay," said Robert, and he went at it with great gusto, curious about what would come up on the screen:

$$1 \div 1 = 1$$
$$2 \div 1 = 2$$
$$3 \div 2 = 1,5$$
$$5 \div 3 = 1,6666666666 \cdots$$
$$8 \div 5 = 1,6$$
$$13 \div 8 = 1,625$$
$$21 \div 13 = 1,615384615 \cdots$$
$$34 \div 21 = 1,619047619 \cdots$$
$$55 \div 34 = 1,617647059 \cdots$$
$$89 \div 55 = 1,618181818 \cdots$$

"Wild!" he said. "Another pile of numbers that refuse to stop. An eighteen biting its own tail. And a few others looking as unreasonable as they come."

"True, true," said the number devil, "but what else do you see?"

Robert thought for a while and said, "All the numbers—they seem kind of wobbly. The second is bigger than the first, the third smaller than the second, the fourth a little bigger, and so on. They keep swaying from side to side. But the farther we go the less they wobble."

"True. The larger the Bonacci numbers, the more you close in on a very special number:

$$1,618\ 033\ 989 \cdots$$

But don't think that's the end of the story, because the number is one of those unreasonable ones, the kind that go on forever. You can take it as far as you like, but you'll never come to the end of it."

"What else would you expect from Bonacci?" said Robert. "But what I don't understand is why they wobble so around that weird number."

"Oh, that," said the number devil. "That's nothing special. They all do that."

"What do you mean—they *all* do that?"

"I mean they don't necessarily need to be Bonaccis. Let's take two perfectly normal, garden-variety numbers. Tell me the first two that come into your head."

"Seventeen and eleven."

"Good. Now add them together."

"That I can do mentally. Twenty-eight."

"Excellent. Now let me show you on the screen where we go from here."

$$11 + 17 = 28$$
$$17 + 28 = 45$$
$$28 + 45 = 73$$
$$45 + 73 = 118$$
$$73 + 118 = 191$$
$$118 + 191 = 309$$

"Got it," said Robert. "What now?"

"We do precisely what we did with the Bonacci numbers. We divide. Go ahead," the number devil said, "and see what you come up with."

Again the numbers flashed on the screen as Robert entered them:

$$17 : 11 = 1,545\ 454 \cdots$$
$$28 : 17 = 1,647\ 058 \cdots$$
$$45 : 28 = 1,607\ 142 \cdots$$
$$73 : 45 = 1,622\ 222 \cdots$$
$$118 : 73 = 1,616\ 438 \cdots$$
$$191 : 118 = 1,618\ 644 \cdots$$
$$309 : 191 = 1,617\ 801 \cdots$$

"That same crazy number!" Robert cried. "What's going on here? Is it buried in all numbers?"

"It is," said the number devil. "And in nature and in art, if you know how to look. By the way—in case you're interested—let me show you what else 1.618 ⋯ can be."

A monstrous fraction flashed on the screen:

$$1{,}618\cdots = 1 + \cfrac{1}{1 + \cfrac{1}{1 + \cfrac{1}{1 + \cfrac{1}{1 + \cfrac{1}{1 + \cfrac{1}{\cdots}}}}}}$$

"A fraction!" Robert cried. "A fraction that never ever ends, a fraction so grotesque it hurts my eyes to look at it! I hate fractions! Mr. Bockel loves them. He loves to torture us with them. Get that monster out of my sight. Please!"

"Don't panic now. It's just a continued fraction. Though it is amazing that we can entice our 'crazy number' 1.618 ⋯ out of a set of ones that keep shrinking and shrinking, don't you think?"

"I'll think anything you like so long as you spare me fractions. And most of all, fractions that have no end."

"All right, all right. I just wanted to give you a little surprise. Let me try something else if the continued fraction upsets you so. This pentagon, for instance:

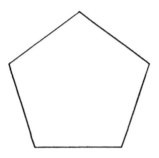

Let's say that each side has the length one."

"One what?" Robert asked immediately. "One meter, one centimeter? Want me to measure it?"

"It doesn't matter," said the number devil, a bit put out. "We had this problem once before, remember? And we agreed to call it one quang. So let's say each side is one quang long. Okay?"

"Okay, okay. Anything you say."

"Now I'm going to draw an orange star inside the pentagon:

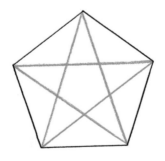

The star consists of five orange lines. Ask me how long any one of those lines is and I'll tell you: precisely 1.618 · · · quang and not one iota more or less."

"That is one weird number!"

"You don't know the half of it," said the number devil with a smile, flattered by Robert's interest. "Now pay close attention. I want you to measure the two orange parts I've marked *A* and *B*."

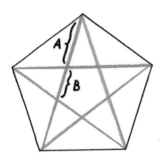 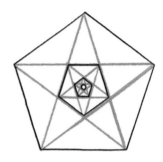

"*A* is slightly longer than *B*," Robert said.

"And just so you don't have to bother your head about it, I can tell you straight away how much longer it is: *A* is precisely 1.618 · · · times longer than *B*. Moreover, we could go on and on—till the cows come home, as you can imagine—because our star is very much like the snowflakes: the orange star has a black pentagon inside it, the black pentagon an orange star inside it, and so on."

"And that blankety-blank unreasonable number keeps turning up?" Robert asked.

"Clearly it does. So if you're not tired of it . . ."

"Not in the least," Robert assured him. "It's fascinating!"

"Then let's go back to your notebook and enter the 'blankety-blank' number. Here, let me dictate:

$$1,618\ 033\ 989\cdots$$

Good. Now subtract 0.5:

$$1,618\ 033\ 989\ldots - 0,5$$
$$= 1,118\ 033\ 989\ldots$$

Double the result. Times two, in other words:

$$1,118\ 033\ 989\ldots \times 2$$
$$= 2,236\ 067\ 978\ldots$$

Good. Now make the new result hop, that is, multiply it by itself. There's a special key for that marked x^2:

$$2,236067977\ldots^2 = 5,000\ 000\ 000$$

"Five!" Robert cried. "No! Impossible! How come? Why five?"

"Well," said the number devil with great pleasure, "we have a five-pointed star inside a five-sided figure."

"Devilishly clever of you," said Robert.

"Now let's make a few dots in our star," continued the number devil. "Place one at every point where the lines cross or come together:

Count how many there are."

"Ten," said Robert.

"And now how many white spaces are there?"

Robert counted eleven.

"Now we need to know how many lines there are. Lines connected by two dots."

It took Robert a while to count them because he kept getting mixed up, but he finally got the answer: twenty.

"Correct," said the number devil. "Now look at this:

$$10 + 11 - 20 = 1$$
$$(D + S - L = 1)$$

If you add the number of dots and spaces together and subtract the number of lines, the total is one."

"So?"

"The thing is, the total is one not only for our star. No, the total is *always* one, no matter what flat figure you start with. It can be as complicated and irregular as you please. Try it. Draw any old figure and you'll see."

He handed Robert the mouse, and the following figures appeared on the screen:

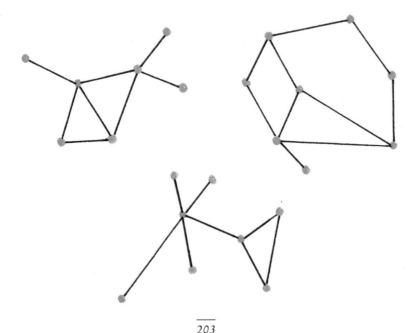

"Don't bother to work it out," said the number devil. "I've done it for you. The first figure has seven dots, two spaces, and eight lines: $7 + 2 - 8 = 1$. The second figure: $8 + 3 - 10 = 1$. The third figure: $8 + 1 - 8 = 1$. Always one.

"By the way, it doesn't work only for flat figures; it works for cubes or pyramids or diamonds as well. The only difference is that then the answer is two rather than one."

"Prove it."

"What you see on the screen is a pyramid."

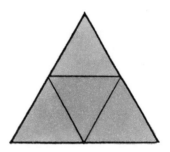

"You call that a pyramid? That's just four triangles."

"But what if you cut and fold it?"

The result flashed immediately on the screen:

"You can do the same with the following figures," said the number devil, drawing three new shapes on the screen:

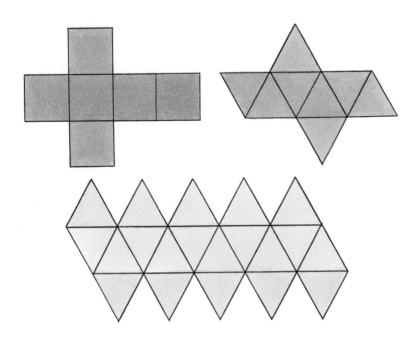

"That's nothing," Robert said. "I've done that before. Cut and paste the first one and you get a cube. But wait a minute, the other two are more complicated."

"Let me show you what you'll get. The second will be a double pyramid pointing both up and down, the third an almost spherical object made of twenty equal-sized triangles:

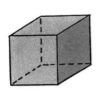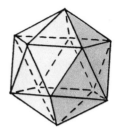

You can even make a kind of ball out of nothing but our favorites, the pentagons. Here's how it looks when you draw it:

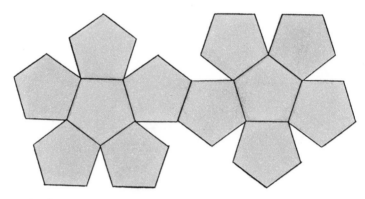

And after the cutting, folding, and pasting:

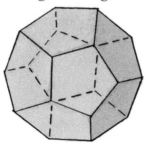

"Not bad," said Robert. "I think I'll make one."
"Not now, please," said the number devil,

"because now I want to get back to our game with the dots, lines, and spaces. Let's start with a cube. It's the easiest."

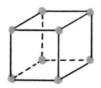

After counting eight dots, six spaces, and twelve lines, Robert said, "8 + 6 − 12 = 2."

"Two, as expected," said the number devil. "No matter what the shape, the result is always two. Dots plus spaces minus lines equals two. No exceptions. Yes, my boy. Anything you can cut and paste. The diamonds on your mother's ring too. And snowflakes, though they always melt before you're through counting . . ."

The number devil's words were quickly growing faint and muffled, the lights in the auditorium fading, and it was beginning to snow on the screen again. This time Robert wasn't scared, however. He knew where he was and knew he wouldn't freeze even though things were turning whiter and whiter before his eyes.

When he woke up, he was lying under a thick white blanket of wool rather than snow, a blanket that had neither dots nor lines nor even what you might call spaces—a blanket that was very definitely four-, not five-sided. And of course the beautiful silver-gray computer had vanished.

What was that blankety-blank number all about, that one point six . . . ? Endless as he knew it was, he could remember no more of it.

Those of you who are handy with scissors and paste may want to try making the figures the number devil showed to Robert. You'll need to draw in little tabs to help you with the pasting. If you do all five and are still game, you can move on to a particularly sophisticated figure—

but only if you are very patient and precise. Take a large sheet of paper (it should be at least 35 × 20 centimeters—that is, 9 × 12 inches—and thick, though cardboard will not work) and copy the figure you see on the next page. Keep in mind that each side of the many triangles must be exactly as long as all the others. You can decide how long you want the sides to be, though three to four centimeters— that is, about an inch and a half (or one quang)—is ideal. Cut out the figure and, using a ruler, fold the paper forward along the red lines and backward along the blue lines. Then paste it together, first the B tab going with the b triangle, the C tab with the c triangle, etc., and finally the A tab with the a triangle. What do you get? A crazy ring made of ten little pyramids. You can turn it forward or back (if you're careful!), and each time you do, a new pentagon and a five-pointed star will come into view. Guess what you get if you count up the dots, spaces, and lines and enter them into our equation:

$$D + S - L = ?$$

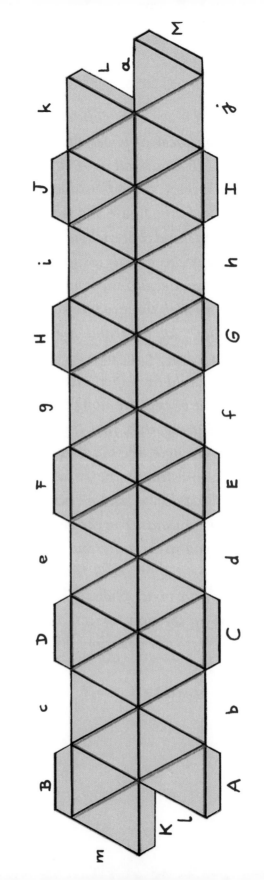

The Eleventh Night

It was nearly dark. Robert was racing though the center of town, though he didn't recognize the streets or buildings. He ran as fast as he could, because Mr. Bockel was after him. Once he was so close that Robert could hear him panting. "Stop!" Mr. Bockel shouted, and Robert spurted forward with all his might. Yet he hadn't the slightest idea why Mr. Bockel was after him or why he was running away from him. Besides, there was no way Mr. Bockel could catch him: the teacher was much too fat.

But when he got to the next corner, what did he see but a second Mr. Bockel tearing out from the left. He bolted across the street, not waiting for the red light to change, and all at once he heard a whole chorus of voices behind him, calling out, "Stop, Robert, stop! We only want to help!"

In addition to the four or five Bockels at his heels and the ones now pouring out of the side streets—

as alike as peas in a pod—there were soon Bockels running straight at him.

Robert shouted for help.

A bony hand grabbed him by the shoulder and pulled him into a doorway. Thank God! It was the number devil.

"Follow me," he whispered. "I know a private elevator that will take us to the top floor."

The elevator was paneled with mirrors, so Robert could see an endless band of number devils and Robert look-alikes.

This is too much, Robert thought, these crowds of the same people!

Still, he was free from the Bockel voices in the street, and when they reached the fiftieth floor and the elevator door opened noiselessly, he and the number devil stepped out into a delightful roof garden.

"You don't know how I've wished for this," Robert said as they rocked peacefully on a swinging bench.

From where they sat, the people down in the street looked like ants.

"I had no idea there were so many Bockels in the world," said Robert.

"Well, they're nothing to you," the number devil said reassuringly. "You needn't be afraid of them."

"*I had no idea there were so many Bockels in the world,*" *said Robert.*
"*Well, you needn't to be afraid of them,*" *the number devil said reassuringly.*

"I guess it's the kind of thing that happens only in dreams," Robert said. "If you hadn't shown up in time, I wouldn't have known what to think."

"That's why I'm here. And now that there's no one to disturb us, tell me what's wrong."

"All week long I've been brooding over what you showed me last time and how it hangs together. I'm glad you showed me all those tricks. They're fun. But I can't help wondering *why*? Why do they turn out the way they do? The blankety-blank number, for instance. And the five. Why do rabbits behave as if they knew what a Bonacci number is? Why don't unreasonable numbers ever end? And why does what you say hold true always and forever?"

"So that's it," said the number devil. "You want to do more than play around with numbers. You want to know what's behind them. The rules of the game, so to speak. The meaning of it all. In other words, you want to know what a mathematician wants to know."

"I don't know what mathematicians want to know. I only know you've *shown* me things, but never *proved* them."

"You're right," said the number devil. "I apologize. The problem is, showing things is easy and, as you put it, fun. Guessing isn't bad, and testing guesses is even better. We've done a lot of that. But none of it is enough. Proof is all. And now you even want hard proof."

"Right," Robert agreed. "Oh, part of what you say I get just like that. But there are things I just don't get at all. I don't see how or why they work."

"To make a long story short, you're dissatisfied. Well, that's good. Do you think we number devils are always satisfied with what we come up with? Not on your life! No, we're constantly contemplating new ways of proving things. Thinking, pondering, meditating—it's a way of life with us. But when a light finally does go on—and it can take ages, because in mathematics the centuries fly by—then we're pleased as punch. Only then are we satisfied."

"You must be exaggerating. It can't be that hard."

"You have no idea!" the number devil replied. "Even when you think you've understood something, you may wake up one morning and realize there's a catch."

"Can you give me an example?"

The number devil rubbed his chin and paused for a moment.

"You think you know everything there is to

know about hopping. What could be simpler than going from 2 to 2×2 and from 2×2 to $2 \times 2 \times 2$?"

"Right: 2^1, 2^2, 2^3, and so on. It's a cinch."

"Yes, but what happens when you hop with zero? 1^0, 8^0, or 100^0? What do you think you get? Shall I tell you? You'll laugh, but the answer is one. Always one:

$$1^0 = 1, \quad 8^0 = 1, \quad 100^0 = 1$$

"How can that be?" Robert was amazed.

"Don't ask. I could prove it to you, but you'd go mad in the process."

"Try me!" said Robert. Now *he* was angry.

But the number devil remained calm.

"Have you ever tried to cross a raging stream?" the number devil asked.

"Have I?" Robert cried. "I'll say I have!"

"You can't swim across: the current would sweep you into the rapids. But there are a few rocks in the middle. So what do you do?"

"I see which ones are close enough together so I can leap from one to the next. If I'm lucky, I make it; if I'm not, I don't."

"That's how it is with mathematical proofs," the number devil told Robert. "But since mathematicians have spent a few thousand years finding ways to cross the stream, you don't need to start from scratch. You've got all kinds of rocks to rely on. They've been tested millions of times and are guaranteed slip-resistant. When you have a new idea, a conjecture, you look for the nearest safe rock, and from there you keep leaping—with the greatest of caution, of course—until you reach the other side, the shore."

"That's all well and good," said Robert, "but tell me, where *is* the shore for numbers or pentagons or hopping?"

"Good question," said the number devil. "The shore is a couple of simple sentences. Couldn't be simpler. Once you get to them, you're home free. They're your proof."

"What are they?"

"Well, here's one: Every ordinary number, be it fourteen or fourteen billion, may be followed by one and only one number, namely, that number plus one. Here's another: A point may not be divided, because it has no area. And yet another: Two points on an even plane may be connected by only one line, which then continues endlessly in both directions."

"You have to look each time you leap," said the number devil. "Sometimes the rocks are so far apart you fall in."

"I see," said Robert. "And starting with a couple of sentences you can leap your way to those blankety-blank numbers or even to the Bonacci numbers?"

"Easily."

The number devil's eyes shone.

"You can go even farther. You just have to look each time you leap. Picture yourself in the middle of the stream. You've got rocks to rely on, remember? Sometimes the rocks are so far apart that you can't make it from one to the next, and if you try jumping, you fall in. Then you have to take tricky detours, and even they may not help in the end. You may come up with an idea, but then you find that it doesn't lead anywhere. Or you may find that your brilliant idea wasn't so brilliant after all."

The number devil looked over at Robert with a gentle smile. "Remember what I showed you the very first time, the numbers I conjured up out of ones?

$$1 \times 1 = 1$$
$$11 \times 11 = 121$$
$$111 \times 111 = 12321$$
$$1111 \times 1111 = 1234321$$

And so on and so forth. It looked as though we could go on indefinitely."

Robert remembered that night well.

"Right, and you got so angry when I tried to say there was something fishy about it. But you know I only said that to get your goat. I had nothing at all to back it up with."

"You had a good nose," the number devil admitted. "Later I went back to it, and you know what? When I came to

$$1\ 111\ 111\ 111 \times 1\ 111\ 111\ 111$$

I fell in! All I got was number hash. So even though the formula worked well enough for a while, it collapsed in the end, without proof.

"In other words, even a number devil can fall on his face. I remember one—the Man in the Moon his name was—who put an idea into a formula he thought would always come out without exception. Well, he tested it a billion five hundred million times—the madman—and each time it worked. He computed himself half to death with his giant computer—he was much, much more precise than we were with our blankety-blank $1.618\cdots$—until he was absolutely positive that the formula would work forever. Then he sat back, satisfied.

"It wasn't long, however, before another number devil—I've forgotten his name—came on the scene, and he computed himself three-quarters to death, and with even greater precision, and what did he find?" The number devil looked at Robert. "That the Man in the Moon was wrong. His wonderful formula worked almost always, but not always. And almost isn't enough. The poor devil! He was working on prima-donna numbers. A ticklish business, let me tell you, and fiendishly difficult to prove."

"I agree," said Robert. "Even when only a few lousy pretzels are at stake. It drives me crazy the way Mr. Bockel goes on about why it takes x hours for y bakers to bake pretzels. I much prefer your tricks."

"You're too hard on him. Think of the poor man knocking out lesson plans night after night. He can't go rock leaping as we do whenever we please.

I feel sorry for him. I bet he's gone home to correct your homework."

Robert looked down at the street, and sure enough there wasn't a single Bockel in sight.

"Many of us," the number devil went on, "have an even harder time of it than your Bockel, however. One of my older colleagues, the well-known Lord Rustle, once took it into his head to prove that $1 + 1 = 2$. Look at this proof. This is how he went about it."

*54·42. $\vdash :: \alpha \in 2 . \supset :. \beta \subset \alpha . !\beta . \beta \neq \alpha . \equiv . \beta \in \iota``\alpha$

Dem.

$-. *54·4. \quad \supset \vdash :: \alpha = \iota`x \cup \iota`y . \supset :.$

$$\beta \subset \alpha . \exists! \beta . \equiv : \beta = \Lambda . v . \beta = \iota`x . v . \beta = \iota`y .$$

$$\underset{v . \beta = \alpha : \exists! \beta}{\frown}$$

$[*24·53·56. *51·161] \qquad \equiv : \beta = \iota`x . v . \beta = \iota`y . v . \beta = \alpha \quad (1)$

$\vdash . *54·25 . \text{Transp} . *52·22 . \supset \vdash : x \neq y . \supset . \iota`x \cup \iota`y$

$$\underset{\neq \iota`x . \iota`x \cup \iota`y \neq \iota}{\frown}$$

$[*13·12] \supset \vdash : \alpha = \iota`x \cup \iota`y . x \neq y . \supset . \alpha \neq \iota`x . \alpha \neq \iota`y \quad (2)$

$\vdash . (1) . (2) . \supset \vdash :: \alpha = \iota`x \cup \iota`y . x \neq y . \supset :.$

$$\beta \subset \alpha . \exists! \beta . \beta \neq \alpha . \equiv : \beta = \iota`x . v . \beta = \iota`y :$$

$[*51·235] \qquad\qquad\qquad\qquad\quad \equiv : (\exists z) . z \in \alpha . \beta = \iota`z :$

$[*37·6] \qquad\qquad\qquad\qquad\qquad\quad \equiv : \beta \in \iota``\alpha \qquad (3)$

$\vdash . (3) . *11·11·35 . *54. 101 . \supset \vdash . \text{Prop.}$

*54·43. $\vdash :. \alpha, \beta \in 1 . \supset : \alpha \cap \beta = \Lambda . \equiv . \alpha \cup \beta \in 2$

Dem.

$\vdash . *54·26 . \supset \vdash :. \alpha = \iota`x . \beta = \iota`y . \supset : \alpha \cup \beta \in 2 . \equiv x \neq y .$

$[*51·231] \qquad\qquad\qquad\qquad\qquad \equiv . \iota`x \cap \iota`y = \Lambda .$

$[*13·12] \qquad\qquad\qquad\qquad\qquad \equiv . \alpha \cap \beta = \Lambda \quad (1)$

$\vdash . (1) . *11·11·35 . \supset$

$\qquad \vdash :. (\exists x, y) . \alpha = \iota`x . \beta = \iota`y . \supset : \alpha \cup \beta \in 2 .$

$\qquad\qquad\qquad\qquad\qquad\qquad \equiv . \alpha \cap \beta = \Lambda \quad (2)$

$\vdash . (2) . *11·54 . *52·1 . \supset \vdash . \text{Prop.}$

"Ugh!" said Robert with a shiver. "That's horrendous. And all to show that $1+1=2$? Something he knew anyway?"

"That's right." The number devil nodded. "Even though everybody knows that $1 + 1 = 2$, Lord Rustle wanted to prove it. Now you see where that can lead."

The number devil paced up and down, getting carried away.

"By the way," he continued, "there are all sorts of problems that look almost as obvious as $1+1=2$ and are just as hard to prove mathematically. Let's say you're going to take a few weeks to drive around the country and you have twenty-five friends to visit, each in a different city. You get out your map and look for the shortest route to save the most time and money. What's the best way to go about it?

"Sounds easy, doesn't it? Well, let me tell you, a good many people have lost a good deal of sleep over it. Even some of the most crafty number devils. And not one of them has cracked it, not to this day."

"What's so hard about it?" Robert asked, surprised. "You just go through all of the many possibilities, you chart each of them precisely on the map, and then you work out which is the shortest possible route."

"What you mean is you make a kind of grid with twenty-five dots."

"Right. If I wanted to visit only two friends, there'd be only one route, from *A* to *B:*

"Well, actually there are two routes. You could also go from *B* to *A*."

"But that's the same," said Robert. "Now, if I wanted to visit three friends . . ."

"Then there would be six possibilities:

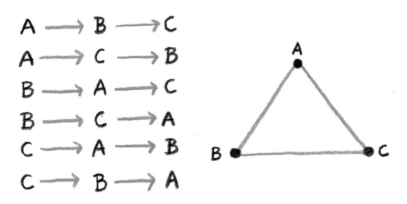

Note that all these routes are equal in length. By the time we have four friends to visit, things are considerably more complex:

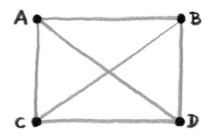

"You know," said Robert, "I'm not quite in the mood to count up all the routes."

"Well, there are precisely twenty-four," said the number devil. "It's very much like the seating problem in your class, I'm afraid. You remember the terrible time we had with Al and Betsy and Charlie and the gang because there were so many possibilities?"

"You bet I remember! Three kids, three vroom! Four kids, four vroom! And so on."

"Well, it's the same with the tour problem."

"Then why is it so hard to solve?" Robert asked. "All I need is to find the various routes and pick the shortest."

"That's what you think!" the number devil replied. "With twenty-five friends you have twenty-five vroom! possibilities, and that's an awfully large number. Approximately

1 600 000 000 000 000 000 000 000 0

You can't possibly test them all and determine which is the shortest. It's beyond even the most powerful computer in the world."

"In other words, no way."

"Not necessarily. We've racked our brains over this one for quite some time now, and, as I say, the cleverest number devils have tried every trick in the book. Sometimes we can work it out and sometimes we can't."

"Too bad," said Robert. "Sometimes is only halfway."

"What's worse, we can't even prove with any certainty that *no* perfect solution exists. That would be something at least. It would allow us to stop looking for one. Besides, proving that no proof exists is a proof in itself of sorts."

"Hm," said Robert. "It's comforting to know that number devils can fall on their faces. I thought you could conjure your way in and out of everything."

"It just looks that way. Many's the time I've been stuck in the middle of the stream. There are times I feel fortunate to get out with dry feet. Heaven knows I don't want to put myself up there with the greats, but even they—and you'll eventually come to know some of them—even they have their problems. Which only means that mathematics will never be over and done with. And a good thing too: there will always be plenty to keep us busy. So good-bye for now, my boy. First thing in the morning I'll be looking into the simplex algorithm for polytopic surfaces."

"The what?"

"The best way to unravel a foul-up. And for that I need a good night's sleep. I'm off to bed. Good night."

And with that, the number devil disappeared, leaving Robert to rock peacefully on the swinging bench.

I wonder what polytopic surfaces are, Robert thought, but then decided it didn't matter. All that mattered was that he didn't have to worry about Mr. Bockel anymore: whenever Mr. Bockel got on his back, the number devil would come to his rescue.

The night was mild and Robert gave himself up to the pleasure of rocking and dreaming, rocking and thinking about nothing at all till the break of day.

The Twelfth Night

Robert had stopped dreaming. There were no giant fish to gobble him up, no ants to crawl up his legs. Even Mr. Bockel and all his look-alikes left him in peace. He slid no more, he froze no more, he was no longer locked away in cellars. He slept better than he had ever slept before.

That was all well and good, but it also got to be rather boring. What was the number devil up to? Maybe he'd had a good idea and couldn't prove it. Or maybe he'd got bogged down in his polyp surfaces (or whatever it was he'd talked about last time).

Though maybe he'd forgotten about Robert.

Robert didn't like that idea in the least.

His mother couldn't understand why he spent hours on end drawing dots and lines and mumbling about visits to nonexistent friends in cities he'd never seen.

"Do your homework, Robert," she would say.

Once Mr. Bockel caught him scribbling away in class.

"What are you up to there, Robert? Give it here."

But Robert had managed to crumple the paper with the huge number triangle into a ball and to sneak it over to Charlie. He could count on good old Charlie: Charlie was used to covering for him.

One night, however, he slept so soundly that it took him a long time to realize there was someone pounding on his door.

"Robert! Robert!"

And who did he see when he finally jumped out of bed and opened it but the number devil.

"I'm so glad to see you," he said. "I've missed you."

"Come quickly!" the number devil said. "I've got an invitation for you! Here!" And he handed him a printed card with a gold border. It said:

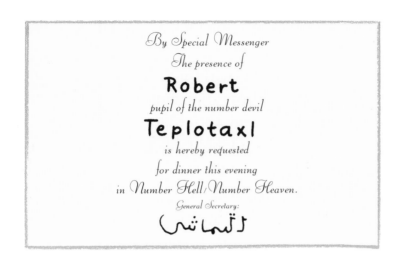

By Special Messenger
The presence of
Robert
pupil of the number devil
Teplotaxl
is hereby requested
for dinner this evening
in Number Hell/Number Heaven.
General Secretary:
كآتاسلاشن

The signature was an illegible squiggle that appeared to be in Arabic script.

"So your name is Teplotaxl," Robert said, pulling on his clothes as fast as he could. "Why didn't you tell me before?"

"Only the inner circle can know a number devil's name," he replied.

"Does that mean I now belong to the inner circle?"

"Apparently. Otherwise you wouldn't have received an invitation."

"Funny," Robert said. "The invitation says 'Number Hell/Number Heaven.' What's that supposed to mean? Either one or the other?"

"Number Paradise, Number Hell, Number Heaven—it's all the same in the end." He went over to the window and opened it as far as it would go. "You'll see. Ready?"

"Yes," said Robert, though he was starting to feel a bit uneasy about the whole thing.

"Then climb on my shoulders."

Robert was afraid he'd be too heavy for the number devil, who was no giant, but he didn't want to insult him. And, lo and behold, no sooner had he put his arms around the number devil's neck than the number devil leaped out of the window and soared into the air.

This could only happen in a dream, Robert thought. But why not? A flight without noise, without safety belts, without silly stewardesses giving you coloring books and plastic toys as if you were three years old—it was good for a change.

It ended in a perfect landing on the terrace of a magnificent palace.

"Here we are," said the number devil, letting Robert down.

"Where's my invitation?" asked Robert. "I'm afraid I left it at home."

"Don't worry," said the number devil. "Anyone who really wants to can enter. The trick is to get here in the first place. And *that*, as you can imagine, very few manage."

No sooner had Robert put his arms around the number devil's neck than the number devil leaped out of the window and soared into the air.

Robert had in fact noticed that nobody seemed to be checking the people going through the giant double door.

So in they went and set off down a long, long corridor with endless doors, many of which were open to varying degrees.

Robert peeked into the very first room. Teplotaxl put his finger to his lips and said, "Sh!" What they saw was a very old man with snow-white hair and a long nose. He was toddling round and round in circles and carrying out a great debate with himself.

"All Englishmen are liars," the man mumbled, "but if *I* say it, what then? I'm an Englishman myself. So I'm lying too. But then what I've just said—namely, that all Englishmen are liars—is not true. But if Englishmen tell the truth, then what I said before must be true as well. In other words, we *are* liars."

The number devil signaled to Robert, and on they went.

"That's poor Lord Rustle," Teplotaxl explained to his guest. "You remember, the one who proved $1 + 1 = 2$."

"Isn't he a bit . . . confused?" Robert asked. "Not that I'm surprised. He must be terribly old."

"He's not the least bit confused; he's got all his

wits about him. As for his age, age means nothing here. Besides, Lord Rustle is one of our youngest. He hasn't reached 150 yet."

"So you have some pretty elderly residents here in the palace."

"You'll see," said Teplotaxl. "In Number Hell— I mean, Number Heaven—no one ever dies."

They came to another open door and peered in to see a man squatting in the corner. He was so tiny that Robert didn't notice him at first, though that may have been because the room was filled with the most curious objects. A few were large pretzels made of glass. Mr. Bockel would have been pleased, thought Robert, though you couldn't eat them and they had the weirdest shapes.

The number devil called Robert's attention to an unusual-looking green bottle.

"Examine it carefully," Teplotaxl whispered into

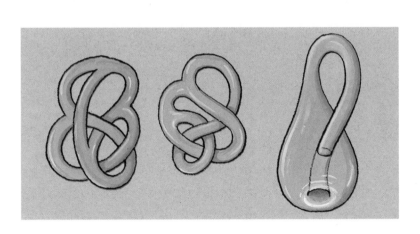

Robert's ear. "Can you tell the inside from the outside?"

Unbelievable! thought Robert. It could only exist in a dream.

"Imagine you wanted to paint the inside blue and the outside red. What would you do? There are no edges. You wouldn't know where to stop the blue or start the red."

"And that little man is the one who invented it?" asked Robert. "He looks as if he would be happier inside the bottle."

"Not so loud! Do you what his name is? Dr. Little. Dr. Happy Little. But let's go now. We have lots more to see."

The next few doors they passed had DO NOT DISTURB signs on them, but one was wide open, and there they stopped. The walls and furniture were coated with a fine dust.

"That's no ordinary dust," said Teplotaxl. "It has more granules than a body could count in a lifetime. If you wanted to cover the head of a pin with it, you'd have to gather all the dust in this room. The man you see there is Professor Singer. He's the man who discovered the dust."

Professor Singer, a pale man with a goatee and piercing eyes, was singing to himself and doing a nervous little dance around the room. "Infinity times infinity is infinity," he sang. "Superinfinity times infinity is superinfinity."

Let's get out of here, Robert thought.

His friend knocked politely on one of the next doors, and a friendly voice called out to them, "Come in!"

Teplotaxl was right: the palace's inhabitants were so old that he seemed a stripling by comparison. Nonetheless the two very old men who greeted them turned out to be quite lively.

"Welcome, gentlemen, welcome," said one of them, a man with large eyes and a wig. "My name is Owl, and this is Professor Horrors."

The latter looked very stern, and he scarcely glanced up from his papers. Robert had the feeling he was less than happy with their visit.

"We were just chatting about the prima-donna numbers," said the friendly one. "A fascinating topic, as I'm sure you are aware."

"Oh, yes," said Robert. "You never know where you stand with them."

"Right you are. But with the help of my colleagues I hope to get to the bottom of them."

"And what does Professor Horrors do, if I may ask?"

But Professor Horrors refused to divulge the subject of his labors.

"Professor Horrors is responsible for a brilliant discovery, a whole new category of number, in fact. Tell our friends here what you call them, will you?"

"Im," said the man with the stern look about him. And that was all he said.

"He means the imaginative numbers," Teplotaxl explained hurriedly, and then he apologized to the eccentric gentlemen for having interrupted their labors.

And so it went. They looked in on Bonacci, but his room was teeming with rabbits; they passed rooms where Mayans and Arabs and Persians and Indians were working and talking and sleeping. The farther they went, the older the people seemed to be.

"The one over there, the one who looks like a maharajah," said Teplotaxl, "he's at least two thousand years old."

The rooms grew in size and splendor until finally Teplotaxl and Robert stood before a kind of temple.

"We're not allowed in here," said Teplotaxl. "The man in white you'll see is so important that a little devil like me can't even say *boo* to him. He's from Greece, and you wouldn't believe the things he's discovered! See those tiles on the floor? The ones with the pentagons and stars? Well, one day he decided to cover the floor with them and when he couldn't do it without leaving gaps between the tiles he came up with unreasonable numbers to explain why. Rutabagas, remember? The rutabaga of two, the rutabaga of five. And the blankety-blank numbers. You remember those, of course."

"Oh, yes," Robert assured him.

"Well, Pythagoras is the man's name," the number devil whispered. "You know what else he came up with? The word *mathematics*! Anyway, here we go."

The hall they now entered was the largest Robert had ever seen. It was bigger than a gym, bigger than a cathedral, and much, much more beautiful. The walls were decorated with mosaics of the most varied patterns, and a majestic throne of gold stood at the first landing on a gigantic flight of steps that led so high there was no telling where it ended.

Robert couldn't get over it. He had never dreamed the number devil lived in such luxury.

"Number Hell, my foot! This is Paradise!"

"Don't be so sure. Oh, I can't complain, but there are times late at night when I'm getting nowhere with my problem and I think I'll go out of my mind! I'm only one step away from the solution but a wall has grown up to keep me from it. That's hell!"

Robert tactfully held his tongue and looked around. Only then did he notice the endlessly long table in the middle of the hall and the waiters along the walls. Suddenly a beanpole of a man at the entrance swung a stick as far back as it would go and struck a gong, which resounded all through the palace.

"Follow me," said Teplotaxl. "Our places are at this end of the table."

Once they had taken their seats, they watched the more famous number devils file past. Robert recognized the Owl and Professor Horrors and then Bonacci (from the rabbit on his shoulder), but most of them he had never seen before. There were solemn Egyptians, there were Indians with pink dots on their foreheads, there were Arabs wearing burnooses and monks in habits, there were Africans and American Indians, Turks with curved swords, Americans in jeans. There were thousands of them.

Robert was amazed at how many number devils there were, but also at how few women he saw among them. He spotted no more than six or seven, and no one seemed to be taking them seriously.

"Why aren't there more women?" he asked. "Is there any rule against them?"

"They used to have a hard time of it. Palace policy was clear: Mathematics is man's work. But things seem to be changing of late."

When the guests had finally taken their seats and mumbled their greetings, the beanpole struck his gong again and the hall fell silent. A Chinese gentleman in fine silk raiment ascended first the stairs and then the throne.

"Who is that?" Robert asked.

"Could be the man who invented zero," Teplotaxl whispered.

"Is he the greatest of them all?"

"The second-greatest," said Teplotaxl. "The greatest lives up where the stairs lead, in the clouds."

"Is he Chinese too?"

"Nobody knows. None of us has ever seen him face-to-face. But we revere him greatly. He is commander-in-chief of all number devils, the man who discovered one. Though for all we know he may not be a man at all. He may be a woman!"

Robert was so impressed that he could say nothing.

Meanwhile the waiters had begun to serve the meal.

"Hey, they're starting with dessert!" Robert said, when a waiter placed a slice of pie on his plate.

"Sh! Not so loud, my boy. We eat nothing but pies, because pies are round and the circle is the most perfect of all figures. Here, try one."

Robert had never tasted anything so delicious in all his life.

"Supposing you want to find out how big the pie is," the number devil said. "How would you go about it?"

"Don't know. You've never shown me how and in school we're still on pretzels."

"What you need is an unreasonable number, the most important of them all. A man all the way up at the head of the table discovered it more than two thousand years ago. Another Greek. Without it we wouldn't know, to this day, how big a pie like this is—or how big our wheels or rings or oil tanks are. In other words, anything that's round. Even the moon and our very own earth. Without the number *pi* we'd be lost."

By this time the room was abuzz with number

devils having a good time, though here and there Robert saw one staring into space and another one making balls out of pie dough. Otherwise they all ate heartily and drank heartily (from pentagonal crystal glasses, fortunately, and not Dr. Little's weird bottles).

When the repast was over, the beanpole sounded the gong again and the man who may have discovered zero rose from his throne and disappeared up the stairs. Gradually the other number devils stood as well—the most eminent first, of course—and started back to their rooms. Soon only Robert and his protector were left.

Just as they were about to go, a gentleman in a magnificent uniform came up to them.

He must be the General Secretary, Robert thought, the man who signed his invitation.

"So this is your apprentice," said the dignitary in a sober voice. "Rather young, don't you think? He hasn't done any conjuring on his own yet, has he?"

"Not yet," Robert's friend replied. "But it won't be long now. Not at the rate he's going."

"How is he doing with prima-donna numbers? Does he know how many there are?"

"Precisely as many as there are ordinary and odd and hopping numbers," said Robert quickly.

"Very good. He can skip the examination. What is his name?"

"Robert."

"Stand, Robert. By the power vested in me as General Secretary, I hereby accept you into the lowest rank of number apprentices and bestow upon you in recognition thereof the Order of Pythagoras, Fifth Class."

With these words he ceremoniously hung a heavy golden chain with a five-pointed star around Robert's neck.

"Thank you," said Robert.

"It goes without saying that this distinction shall remain a secret," the General Secretary added and, without so much as a nod, turned on his heel and disappeared.

"Well, that's that," said Robert's friend and master. "I must be going. You're on your own now."

"What?" cried Robert. "You can't leave me like that!"

"Sorry," said Teplotaxl, "but I have my own work to attend to."

Robert saw that Teplotaxl was moved. Robert was too. On the brink of tears, in fact: he hadn't realized how much a part of his life the number devil had become. But neither the one nor the other felt it seemly to show his emotions, so all Teplotaxl said was "Good-bye, Robert," and all Robert said was "Bye."

In a twinkling his friend was gone.

Robert was now all alone at the gigantic table. How in heaven's name was he going to get home? he wondered. He felt the chain weighing heavier around his neck and the delicious pie growing heavier in his stomach, and before long he had nodded off and was soon so fast asleep that for all he knew he had never left his room on his master's shoulders.

He awoke in his bed, of course, with his mother shaking him and saying, "Time to get up, Robert. If you don't get up this very instant you'll be late for school."

The same as always, Robert said to himself. In your dreams you get delicious pies to eat, you may even get a chain with a star hung around your neck, and the minute you open your eyes everything's back to normal.

But as he stood in front of the mirror in his pajamas, brushing his teeth, he felt something tickling his chest and looked down to see a tiny five-pointed star on a thin golden chain. He couldn't believe his eyes. This time his dream had come true!

After he'd dressed, he took the chain off and stuck it in his pocket: he didn't want his mother asking silly questions.

Where did that star come from? she'd want to know the minute she saw it. Boys don't wear jewelry!

How could he tell her it was the emblem of a secret order?

Things were normal in school except that Mr. Bockel looked more tired than usual and almost immediately took cover behind the newspaper to eat his pretzels in peace. But first he set forth a problem he was sure would take the whole hour to solve.

"How many of you are there in the class?" he asked, and the eager-beaver Doris shot up her hand and said, "Thirty-eight."

"Good, Doris. Now I want you to pay close attention. If the first boy down here in the front—

what's your name again? oh yes, Albert— well, if Albert gets one pretzel, and Betsy gets two pretzels, and Charlie three, Doris four, Enrique five, Felice six, and so on all the way up to number thirty-eight back there, how many pretzels do I need to buy to supply the entire class?"

Robert was incensed.

Mr. Bockel and his mind-bockeling problems! He thought, Now we'll have to work like the devil while he takes it easy.

Robert didn't let on how annoyed he was at the situation, but while all his classmates hunched over their notebooks he simply stared out into thin air.

"What's the matter, Robert?" Mr. Bockel asked. "Dreaming again?"

So he does keep one eye on the class, said Robert to himself.

"No, no," he said aloud. "I'm working on your problem." And he started writing:

$$1 + 2 + 3 + 4 + 5 + 6 \cdots$$

God, how boring! By the time he got to eleven he was totally confused. He, the bearer of the

Order of Pythagoras (even if it was only fifth class)! But then he realized he wasn't wearing the star. He'd left it in his pocket.

Cautiously—he didn't want Mr. Bockel to see—he hung it around his neck under his shirt, and in a flash he had the most elegant solution to the problem. Triangle numbers. What else! It was as if he had learned them just for the occasion. How did they go now?

He wrote the following:

$$\begin{array}{cccccc} 1 & 2 & 3 & 4 & 5 & 6 \\ 12 & 11 & 10 & 9 & 8 & 7 \\ \hline 13 & 13 & 13 & 13 & 13 & 13 \end{array}$$

$$6 \times 13 = 78$$

If it worked for the numbers from one to twelve, it would work for one to thirty-eight!

$$\begin{array}{cccccc} 1 & 2 & 3 & \cdots & 18 & 19 \\ 38 & 37 & 36 & \cdots & 21 & 20 \\ \hline 39 & 39 & 39 & \cdots & 39 & 39 \end{array}$$

$$19 \times 39 = \text{?}$$

Cautiously again he pulled the calculator out of his briefcase and entered the following under the desk:

$$19 \times 39 = 741$$

"Got it!" he cried. "Nothing to it!"

"What?" said Mr. Bockel, letting his newspaper drop. "Well, what's the answer?"

"741," said Robert softly.

You could have heard a pin drop.

"How did you do it?" Mr. Bockel asked.

"Oh," said Robert, clutching the star under his shirt, "it practically solved itself." And he silently thanked the number devil for all he had done for him.

Warning!

Nothing in a dream is quite as it is in school or books. When Robert and the number devil talk about mathematics, they use some unusual expressions. No wonder. *The Number Devil* is anything but a usual story.

So don't think the dream words they use will be understood by everyone. Your mathematics teacher, for instance, or your parents. Mention "hopping" or "rutabagas" to them, and they won't know what you're talking about. Grownups have different words for such things: when they make a number hop twice, they square it or raise it to the power of two; they don't take the rutabaga, they take the square root. In mathematics classes "prima-donna numbers" are called prime numbers, "imaginative numbers" imaginary numbers, and you will never hear your teachers say "five vroom!" because the way they learned to say it from their teachers is five factorial.

Technical terms don't exist in dreams. Nobody dreams in big words. So when the number devil talks in images and sets numbers hopping instead of raising them to power of x or y, it's not just kid stuff. In dreams we do as we please.

In class, however, we never sleep and seldom dream. Your teacher is right to use the expressions used by mathematicians the world over. Please do the same (or you may get into trouble).

Seek-and-Ye-Shall-Find List

Once you've finished *The Number Devil,* you may not remember whether something you need is in it. The following list will help you to locate it if it is.

The list contains the dream words used by Robert and the number devil as well as the "correct," "official" terms, the ones mathematicians use. The latter appear in ordinary type, the dream words in *italics.*

The list also contains several expressions not used in the book at all. You don't need to worry about them. I have included them in case *The Number Devil* falls into the hands of mathematics teachers or other grownups (who deserve a chuckle of their own, after all).

Acknowledgments

Since the author is by no stretch of the imagination a mathematician, he is much obliged to those who have given him a helping hand.

First and foremost he would like to thank his mathematics teacher Theo Renner (a pupil of Sommerfeld's), who, unlike Mr. Bockel, was living proof that mathematics is a matter of pleasure rather than fear.

Among more recent number devils the following have done work I found particularly useful: John H. Conway, Philip J. Davis, Keith Devlin, Ivar Ekeland, Richard K. Guy, Reuben Hersh, Konrad Jacobs, Theo Kempermann, Imre Lakatos, Benoît Mandelbrot, Heinz-Otto Peitgen, and Ian Stewart. Pieter Moree of the Max Planck Institute for Mathematics in Bonn was kind enough to read through the text and correct a few errors.

It goes without saying that none of the forenamed gentlemen is responsible for Robert's dreams.

Munich, Autumn 1996   H.M.E.